Urban Agglomeration - Extracting Lessons for Sustainable Development

Edited by Rui Alexandre Castanho

Published in London, United Kingdom

Urban Agglomeration - Extracting Lessons for Sustainable Development
http://dx.doi.org/10.5772/intechopen.1000332
Edited by Rui Alexandre Castanho

Contributors
Abdulkarim Hasan Rashed, Ana Vulevic, Gualter Couto, Kateryna Vasiutynska, Levine Daphna, Mara Franco, Rifaat Abdalla, Rui Alexandre Castanho, Sergey Barbashev, Stabak Roy, Tomoko Kubo, Yoshimichi Yui

First published in London, United Kingdom, 2024 by IntechOpen
IntechOpen is the global imprint of INTECHOPEN LIMITED, registered in England and Wales, registration number: 11086078, 167-169 Great Portland Street, London, W1W 5PF, United Kingdom

British Library Cataloguing-in-Publication Data
A catalogue record for this book is available from the British Library

Additional hard and PDF copies can be obtained from orders@intechopen.com

Urban Agglomeration - Extracting Lessons for Sustainable Development
Edited by Rui Alexandre Castanho
p. cm.
Print ISBN 978-0-85466-686-7
Online ISBN 978-0-85466-685-0
eBook (PDF) ISBN 978-0-85466-687-4

For EU product safety concerns:
IN TECH d.o.o., Prolaz Marije Krucifikse Kozulić 3, 51000 Rijeka, Croatia,
info@intechopen.com or visit our website at intechopen.com.

Meet the editor

Rui Alexandre Castanho holds an international Ph.D. in Sustainable Planning in Borderlands. He is Vice-Dean for Sustainable Development, WSB University, Poland, and a visiting professor at the University of Johannesburg, South Africa. He is also a European Climate Pact Ambassador. Dr. Castanho completed a postdoctoral research appointment on the GREAT Project, University of Azores, Portugal. He is the vice-coordinator of the VALORIZA - Research Centre for Endogenous Resource, Polytechnic Institute of Portalegre (IPP), Portugal, and collaborates with the Centre for Tourism Research, Development and Innovation (CITUR), Portugal, and the AQUAGEO Research Group, University of Campinas (UNICAMP), Brazil. Dr. Castanho was awarded the title of Doctor Honoris Causa by the Peruvian University of Science and Informatics (UPCI).

Contents

Preface

Urban Agglomeration – Extracting Lessons for Sustainable Development presents a comprehensive overview of the opportunities and challenges of rapid urbanization. As cities grow, urban agglomerations, which are dense hubs of economic, social, and environmental activity, become focal points in the quest for sustainable development. While these urban centers drive innovation, economic growth, and cultural exchange, they also grapple with pressing issues such as resource depletion, environmental degradation, and social inequality. Thus, understanding how to harness the positive aspects of urban agglomeration while mitigating its negative impacts is essential for shaping future resilient and sustainable cities.

By examining global case studies, best practices, and innovative solutions, this book provides valuable lessons that inform policy, planning, and community engagement. At the same time, it delves deeper into the dynamics of urbanization, the interconnection between economic and environmental factors, and the critical role of governance in managing urban spaces effectively.

Urban Agglomeration – Extracting Lessons for Sustainable Development includes seven chapters that highlight the importance of integrated planning, inclusive policies, and technological innovation in addressing the multifaceted challenges of urban agglomerations. It is an informative resource for urban planners, policymakers, academics, and anyone interested in the future of our cities.

Rui Alexandre Castanho
VALORIZA - Research Center for the Enhancement of Endogenous Resources,
Polytehnic Institute of Portalegre (IPP),
Portalegre, Portugal

WSB University,
Dabrowa Górnicza, Poland

Section 1

Sustainable Development and Urbanization, General Overview

Chapter 1

Introductory Chapter: Urban Agglomeration – Extracting Lessons for Sustainable Development – The Opening Act

Rui Alexandre Castanho

1. Introduction

In the tapestry of human civilization, cities have emerged as vibrant threads weaving together the aspirations, innovations, and challenges of our ever-evolving societies. As our planet undergoes unprecedented urbanization, the confluence of people, ideas, and resources in urban agglomerations becomes a focal point for the pursuit of sustainable development [1–10]. The book "Urban Agglomeration: Extracting Lessons for Sustainable Development" embarks on a journey through the intricate landscapes of cities, exploring the dynamic interplay between urbanization and sustainability.

Within the cacophony of skyscrapers and bustling streets lies a narrative of profound significance that unfolds in this book's pages. By delving into the heart of urban agglomerations, we aim to decipher the lessons embedded in their growth, evolution, and adaptation. In doing so, we aspire to guide policymakers, scholars, and citizens toward a more sustainable future.

As we navigate the complexities of urban life, we encounter challenges ranging from environmental degradation to social inequality. However, these challenges lie opportunities for transformative solutions that redefine the urban experience and forge a path toward sustainability.

The following chapters will traverse a diverse range of urban landscapes, examining case studies from metropolises across the globe. Through a multidisciplinary lens, we will analyze urban agglomerations' environmental, economic, and social dimensions. By distilling the successes and setbacks of various cities, we aim to distill principles and strategies that can guide the sustainable development of urban areas in the twenty-first century.

As we embark on this monograph, let us unravel the complexities of urban agglomeration and discern the threads that weave together the fabric of sustainable urban living. Contextually, we strive not only to understand the challenges of today but also to envision the possibilities of tomorrow. Welcome to a journey where cities become laboratories for sustainable development, and the lessons extracted form the blueprint for a resilient and thriving future.

2. Urban agglomerations and sustainable development: a summarized framework

Urban agglomeration, the clustering of population and activities in urban areas, is a pivotal phenomenon shaping the modern world [11–13]. This framework aims to distill key insights and principles for understanding and fostering sustainable development within urban agglomerations. In this regard, it is possible to highlight the following topics:

i. *Dynamics of urbanization*: explore the historical context and drivers of urbanization; examine patterns of population growth, migration, and urban sprawl; or, understand the impact of technological advancements on urban development [14–16].

ii. *Environmental sustainability*: assess the ecological footprint of urban areas; analyze strategies for mitigating environmental degradation; and, explore the role of green infrastructure, renewable energy, and waste management [17–20].

iii. *Economic resilience*: examine the economic engines driving urban agglomerations; assess the role of innovation, entrepreneurship, and industry diversification; and, explore inclusive economic development to address disparities [21, 22].

iv. *Social inclusivity*: investigate the social fabric of urban communities; address issues of housing, healthcare, and education; and, analyze the impact of social policies on inclusivity [23].

v. *Infrastructure and mobility*: evaluate the efficiency and sustainability of urban infrastructure; explore innovative transportation solutions for reducing congestion; and, assess the role of innovative city technologies in enhancing urban mobility [24–28].

vi. *Governance and planning*: examine the role of effective governance in managing urban growth; explore urban planning strategies for resilience and adaptability; and, address challenges related to corruption, bureaucracy, and citizen participation [2, 29].

vii. *Cultural and historical perspectives*: consider the cultural identity and heritage of urban agglomerations; explore the role of cultural preservation in sustainable development; and, analyze the impact of urban development on local traditions and communities [30–34].

viii. *Lessons from global case studies*: showcase successful and innovative urban development initiatives; extract lessons from diverse cities worldwide; and, provide a comparative analysis of strategies employed in different urban contexts [35].

In fact, the above topics—as findings and principles for sustainable urban development—highlight the interconnected nature of environmental, economic, and social factors. Encourage ongoing research and collaboration to refine and implement the framework in diverse urban contexts.

Thereby, this summarized framework guides policymakers, urban planners, and researchers seeking to navigate the complexities of urban agglomeration while fostering sustainability and resilience in our evolving urban landscapes [36–38].

3. Closing act

Contextually, as we conclude our exploration of urban agglomeration and its profound implications for sustainable development, we find ourselves at the crossroads of challenges and opportunities. The urban tapestry, woven with the threads of millions of lives, aspirations, and innovations, beckons us to reflect on the lessons learned and the path ahead.

In traversing the diverse landscapes of urbanization, we have witnessed the dynamic interplay of environmental stewardship, economic vitality, and social inclusivity. The framework provided serves as a compass, guiding us through the complexities of urban agglomerations and offering insights into the multifaceted dimensions of sustainable development.

Nevertheless, our journey is far from over. The urban landscape continues to evolve, presenting new challenges and opportunities that demand our collective attention and ingenuity. As we stand on the precipice of an urbanized future, let us carry forward the following reflections:

a. *Interconnected realities*: recognize the interconnected nature of urban development's environmental, economic, and social factors; and, embrace holistic approaches that address the complexity of urban agglomeration.

b. *Innovation and adaptability*: champion innovation as a catalyst for sustainable urban solutions; and, embrace adaptability in the face of evolving challenges, fostering resilient urban environments.

c. *Inclusivity and equity*: prioritize social inclusivity as a cornerstone of sustainable urban development; and, address disparities in access to resources, opportunities, and services.

d. *Global collaboration*: foster international collaboration and knowledge exchange to address shared urban challenges; and, learn from diverse case studies, recognizing the global significance of local actions.

e. *Empowering governance*: advocate for transparent, effective governance to navigate the complexities of urban growth; and, empower local communities and stakeholders to actively participate in shaping their urban futures.

As we bid farewell to this exploration of urban agglomeration, we should use the insights gained and lessons extracted into the realm of actionable change. With its towering structures and bustling thoroughfares, the cityscape is a canvas awaiting the brushstrokes of sustainable innovation and thoughtful governance.

In the spirit of collaboration, let us work hand in hand—urban planners, policymakers, researchers, and citizens alike—toward weaving a sustainable urban tapestry that stands resilient against the winds of change. Our collective endeavors hold the potential to transform urban agglomerations into beacons of sustainability, where the threads of progress intertwine harmoniously with the fabric of our shared human experience.

Acknowledgements

This study is a result of the National Funds provided by FCT—Foundation for Science and Technology to VALORIZA—Research Center for Endogenous Resource Valorization (project UIDB/05064/2020).

Author details

Rui Alexandre Castanho[1,2]

1 VALORIZA – Research Center for the Enhancement of Endogenous Resources, Polytechnic Institute of Portalegre (IPP), Portugal

2 WSB University, Dabrowa Górnicza, Poland

*Address all correspondence to: acastanho@wsb.edu.pl

References

[1] Deng H, Liu K. Spatiotemporal evolution of urban resilience and spatial spillover effects in Guangdong Province, China. Land. 2023;**12**:1800. DOI: 10.3390/land12091800

[2] Loures L, Naranjo Gómez JM, Castanho RA, Loures A. Benefits and limitations of public involvement processes in landscape redevelopment projects—Learning from practice. In: Regional Intelligence Spatial Analysis and Anthropogenic Regional Challenges in the Digital Age. Basel, Switzerland: Springer; 2020. pp. 29-48. DOI: 10.1007/978-3-030-36479-3_3

[3] Pena D, Albarran A, Lopez-Pineiro A, Rato-Nunes JM, Sanchez-Llerena J, Becerra D. Impact of oiled and de-oiled olive mill waste amendments on the sorption, leaching, and persistence of S-metolachlor in a calcareous clay soil. Journal of Environmental Science and Health – Part B Pesticides, Food Contaminants, and Agricultural Wastes. 2013;**48**(9):767-775

[4] Da Gama JT, Nunes JR, Loures L, Pineiro AL, Vivas P. Assessing spatial and temporal variability for some edaphic characteristics of Mediterranean rainfed and irrigated soils. Agronomy. 2019;**9**(3):132

[5] Ferreira P, Loures L, Nunes J, Brito P. Are renewable energy stocks a possibility to diversify portfolios considering an environmentally friendly approach? The view of DCCA correlation coefficient. Physica A: Statistical Mechanics and Its Applications. 2018;**512**:675-681

[6] Castanho RA, Naranjo Gómez JM, Vulevic A, Couto G. The land-use change dynamics based on the CORINE data in the period 1990-2018 in the European Archipelagos of the Macaronesia Region: Azores, Canary Islands and Madeira. ISPRS International Journal of Geo-Inf. 2021;**10**:342

[7] Kim J, Jin H-Y. Interpreting tactical urbanism through innovation–diffusion theory: Insights from a collaborative design studio experience. Land. 2024;**13**:14. DOI: 10.3390/land13010014

[8] Canesi R, Gallo B. Risk assessment in sustainable infrastructure development projects: A tool for mitigating cost overruns. Land. 2024;**13**:41. DOI: 10.3390/land13010041

[9] Liang L, Chen Z, Chen S, Zheng X. Evaluation of site suitability for photovoltaic power plants in the Beijing–Tianjin–Hebei Region of China using a combined weighting method. Land. 2024;**13**:40. DOI: 10.3390/land13010040

[10] Zhi J, Cao X, Liu W, Sun Y, Xu D, Da C, et al. Remote sensing monitoring and spatial pattern analysis of non-grain production of cultivated land in Anhui Province, China. Land. 2023;**12**:1497. DOI: 10.3390/land12081497

[11] Yu H, Liu D, Zhang C, Yu L, Yang B, Qiao S, et al. Research on spatial–temporal characteristics and driving factors of urban development intensity for Pearl River Delta Region based on Geodetector. Land. 2023;**12**:1673. DOI: 10.3390/land12091673

[12] Loures L, Castanho RA, Vulevic A, Naranjo Gómez J, Cabezas J, Fernández-Pozo L. The multi-variated effect of city cooperation in land use planning and decision-making processes - A European analysis. In: Urban Agglomerations. London, United Kingdom: InTech; 2018. pp. 87-106

[13] Jiao G, Lu L, Chen G, Huang Z, Cirella GT, Yang X. Spatiotemporal characteristics and influencing factors of tourism Revenue in the Yangtze River Delta Urban Agglomeration Region during 2001-2019. Sustainability. 2021;**13**:3658. DOI: 10.3390/su13073658

[14] Fadigas L. Fundamentos Ambientais do Ordenamento do Território e da Paisagem. Lisboa; 2007

[15] Fadigas L. Urbanismo e Território – As políticas públicas. Lisboa; 2015

[16] De Sousa C. Turning brownfields into green space in the City of Toronto. Landscape and Urban Planning. 2003;**62**:181-198

[17] Kaletová T, Loures L, Castanho R. Proposal of Indicators of Ecosystem Services (ES) Provided by Agriculture and Intermittent Rivers and Ephemeral Streams (IRES), Case Study Caia River Basin, Portugal. In: ORDENACIÓN DEL ESPACIO: CIUDADES INTELIGENTES, TURISMO Y LOGÍSTICA. MD, USA: National Library of Medicine; 2018. pp. 293-303

[18] Kaletová T, Loures L, Castanho RA, Aydin E, Gama J, Loures A, et al. Relevance of Intermittent Rivers and streams in agricultural landscape and their impact on provided ecosystem services—A Mediterranean case study. International Journal of Environmental Research and Public Health (IJEPRH). 2019;**16**:2693, 3-2616. DOI: 10.3390/ijerph16152693

[19] Loures L, Panagopoulos T, Burley J. Assessing user preferences on post-industrial redevelopment. Environment and Planning B: Planning and Design. 2016;**43**(5):871-892

[20] Castanho RA, Cabezas J, Naranjo Gómez J, Martín Gallardo J, Fernández-Pozo L, Loures L, et al. Assessing ecosystem services delivered by public green spaces in European Major Cities. In: Landscape Architecture. Thousand Oaks, CA: Intech; 2020. DOI: 10.5772/intechopen.91415

[21] Yilmaz Genç S, Castanho RA, Syed H. Collaborative spatial planning for sustainable growth in the European Union. In: Studies on Social Sciences. Ankara: Publishing House; 2019. pp. 183-206

[22] Jurado Almonte J, Pazos-Garcia F, Castanho RA. Eurocities of the Iberian Borderland: A second generation of border cooperation structures. An analysis of their development strategies. Sustainability. 2020, 2020;**12**(16):6438. DOI: 10.3390/su12166438

[23] Vulevic A, Djordjevic D, Castanho RA, Cabezas-Fernandez J. Social marketing and their related challenges for the limited access for people living with a disability: A Serbian Case Study. In: Case Studies on Social Marketing. Cham, Switzerland: Springer; 2019. DOI: 10.1007/978-3-030-04843_19

[24] Naranjo Gómez J, Castanho RA, Cabezas J, Loures L. Assessment of high-speed rail service coverage in municipalities of Peninsular Spain. Infrastructures. 2020;**2020**(5):11. DOI: 10.3390/infrastructures5020011

[25] Castanho RA, Behradfar A, Vulevic A, Naranjo Gómez J. Analyzing transportation sustainability in the Canary Islands Archipelago. Infrastructures. 2020;**2020**(5):58. DOI: 10.3390/infrastructures5070058

[26] Castanho RA, Naranjo Gómez JM, Vulevic A, Behradfar A, Couto G. Assessing transportation patterns in the azores Archipelago. Infrastructures. 2021;**2021**(6):10. DOI: 10.3390/infrastructures6010010

[27] Couto G, Castanho RA, Pimentel P, Carvalho C, Sousa Á, Graça Batista M, et al. Transportation and Infrastructures' Sustainability in Ultra-peripheral Territories: Studies Over the Azores Region. In: Rocha Á et al., editors. Trends and Applications in Information Systems and Technologies (WorldCIST 2021, AISC 1367). Switzerland: Springer Nature; 2021. pp. 371-379. DOI: 10.1007/978-3-030-72660-7_36

[28] Naranjo Gómez JM, Castanho RA, Vulevic A. Analyzing transportation logistics and infrastructures' sustainability in the Iberian Peninsula: The case of Portugal Mainland. European Planning Studies. 2021;**30**(12):2514-2536. DOI: 10.1080/09654313.2021.2014789

[29] Castanho RA. Identifying processes of smart planning, governance and management in European Border Cities. Learning from City-to-City Cooperation (C2C). Sustainability. 2019;**11**:5476. DOI: 10.3390/su11195476

[30] Mahony K, Zyl J. The impacts of tourism investment on rural communities: Three case studies in South Africa. Development Southern Africa. 2002;**19**(1):83-103. DOI: 10.1080/03768350220123891

[31] Pilving T, Kull T, Suškevics M, Viira AH. Creating shared collaborative tourism identity in a post-communist environment. Scandinavian Journal of Hospitality and Tourism. 2021;**21**(3):313-340

[32] Castanho RA, Couto G, Pimentel. Principles of sustainable tourism and cultural management in rural and ultra-peripheral territories: Extracting guidelines for its application in the Azores Archipelago. Cultural Management: Science and Education (CMSE). 2020;**4**(1):9-24. DOI: 10.30819/cmse.4-1.01

[33] Castanho RA, Santos C, Couto G. Creative tourism in Islands and regional sustainable development: What can we learn from the pilot projects implemented in the Azores territory? Land

[34] Williams A. Introduction. In: Williams A, editor. Southern Europe Transformed-Political and Economic Change in Greece, Italy, Portugal and Spain. New York: Harper and Row; 1984. pp. 1-32

[35] Behradfar A, Castanho RA, Khanian M, Mohammadi S, Mohammadi H, Loures A, et al. Using a study of the social challenges of urbanization in the next 30 years based on age transition and comprehensive plans for Iranian cities: The Case of Hamedan City. WSEAS Transactions on Environment and Development. 2022;**18**:1198-1207. DOI: 10.37394/232015.2022.18.112

[36] Castanho RA, Couto G, Naranjo Gómez JM, Pimentel P, Carvalho C, Sousa Á, et al. Evolutionary dynamics in Azorean Landscapes: The land-use changes in forests and semi-natural areas in the Archipelago from 1990 to 2018. In: Rocha Á et al., editors. Trends and Applications in Information Systems and Technologies (WorldCIST 2021, AISC 1367). Switzerland: Springer Nature; 2021. pp. 244-252. DOI: 10.1007/978-3-030-72651-5_24

[37] Bilalli B, Castanho RA. Sustainable development and renewable energy sources in Turkey. In: Online International Congress of Energy, Economy and Security (ENSCON´20), 14-15 November 2020, Istanbul, Turkey. 2020. p. 46

[38] Portney K. Taking Sustainable Cities Seriously: Economic Development, the Environment, and Quality of Life in American Cities. Cambridge: MIT Press; 2003

Chapter 2

Framework for Assessing the Impacts of Climate Change on Urban Agglomerations: A GIS and Remote Sensing Perspective

Rifaat Abdalla

Abstract

As the specter of climate change looms over urban agglomerations, this concept chapter delves into the transformative potential of GIS and Remote Sensing techniques in dissecting and mitigating its impacts. By intricately analyzing land-cover and surface temperature data, we unveil the nuanced effects of climate change on land surface temperature (LST) across varied land-cover types. Leveraging the expansive spatial coverage of remote sensing data, especially satellite images, we can meticulously monitor urban structures, offering invaluable insights into impervious surfaces and vegetated areas. This trove of information not only enlightens the current state and evolution of urban structures but also becomes the bedrock for effective urban planning strategies and climate change adaptation measures. In tandem, the amalgamation of remote sensing with GIS techniques facilitates a granular exploration of the intra-urban thermal environment and the intricate spatial links between urban vulnerability and characteristics. By delving into these insights, GIS and remote sensing emerge as indispensable allies in the quantification and monitoring of climate change impacts on urban agglomerations, guiding decisive measures for sustainable urban development and climate adaptation.

Keywords: urban agglomerations, climate change, vulnerability, resiliency, disaster management, GIS, remote sensing

1. Introduction

Urban agglomerations are navigating an unprecedented era of challenges under the pervasive influence of climate change. The escalating temperatures, capricious alterations in precipitation patterns, and surges in extreme weather events pose formidable threats to the resilience and sustainability of urban areas worldwide [1]. In response, this chapter advocates the strategic deployment of Geographic Information Systems (GIS) and remote sensing techniques as powerful tools in comprehending and managing the multifaceted impacts of climate change on urban landscapes [2]. Our focus is on the confluence of land-cover and surface temperature data, unraveling the intricate relationships between different land-cover types and the ensuing shifts in land surface temperature (LST) [3]. The arsenal of remote sensing, particularly satellite imagery, comes to the forefront, affording us expansive spatial coverage and recurrently updated information essential for the vigilant monitoring of urban structures [4]. This chapter contends that this dynamic repository of data not only facilitates an understanding of the present urban milieu but becomes an indispensable asset in sculpting effective urban planning strategies and fortifying the adaptive measures against the onslaught of climate change.

The subsequent integration of GIS techniques with remote sensing adds another layer of sophistication to our analytical approach [5]. This union allows us to delve into the complex intra-urban thermal environment, deciphering the spatial variability of the connections between urban vulnerability and characteristics. As we navigate through these methodologies, the overarching aim is to provide a robust framework that quantifies, monitors, and analyzes the impacts of climate change on urban agglomerations [6]. This framework, fortified by GIS and remote sensing, is envisaged as a catalyst for informed decision-making processes, steering urban development toward sustainability and resilience. As we embark on this exploration, it becomes increasingly evident that GIS and remote sensing emerge not merely as technical tools but as indispensable allies in deciphering the intricate dance between climate change and urban agglomerations.

In this context, the integration of Geographic Information Systems (GIS) and remote sensing technologies has emerged as a powerful approach to assess and monitor the effects of climate change on urban landscapes. This amalgamation of advanced technologies forms the bedrock of our exploration, promising a comprehensive understanding of the intricate dynamics between climate change and urban agglomerations, which can be summarized in the following.

1.1 Utilizing land-cover and surface temperature data

GIS and remote sensing stand as pivotal tools in dissecting the impacts of climate change on land surface temperature (LST) under various land-cover types. Leveraging satellite imagery and cutting-edge remote sensing techniques, researchers gain access to robust datasets. This allows for a nuanced assessment of temperature variations across urban landscapes, a crucial insight for comprehending the contribution of different land-cover types to the urban heat island effect. The data becomes instrumental in formulating strategies to mitigate heat-related challenges in urban environments.

1.1.1 Monitoring urban structures

Remote sensing data, with its extensive spatial coverage and frequent updates, facilitates the meticulous monitoring of urban structures. This includes the dynamic tracking of changes in impervious surfaces like roads and buildings, alongside vegetated areas. The integration of GIS enhances this capability, enabling the creation of dynamic maps that vividly illustrate the evolution of urban structures over time. This real-time monitoring becomes a linchpin in supporting urban planning strategies, providing decision-makers with valuable insights into the current state and developmental trajectory of urban areas within municipal boundaries [7].

1.1.2 Supporting Urban Planning Strategies and Climate Change Adaptation

The synergy between GIS and remote sensing data offers indispensable information for supporting urban planning strategies and fortifying climate change adaptation measures. Decision-makers can leverage these tools to identify vulnerable areas, assess the impact of climate change on infrastructure, and devise targeted interventions to bolster urban resilience. The visualization of spatial data empowers planners to make informed decisions, steering the course toward sustainable urban development and the implementation of effective climate change adaptation measures [8].

1.1.3 Assessing intra-urban thermal environment

GIS techniques, in tandem with remote sensing, contribute significantly to the assessment of the intra-urban thermal environment. This involves unraveling the spatial variability of the links between urban vulnerability and characteristics. This nuanced understanding enables a more refined analysis of the impacts of climate change, guiding the development of localized adaptation strategies and the prioritization of resources based on the specific needs of different urban zones [9].

1.2 Research objectives

1.2.1 Spatial analysis of urban Heat Islands

Conduct an intricate spatial analysis utilizing GIS and remote sensing techniques to identify and characterize urban heat islands (UHIs) within different land-cover types. This involves assessing land surface temperature (LST) variations under various urban structures and natural features, providing insights into the intensity and distribution of heat islands.

1.2.2 Temporal monitoring of urban structures

Develop a comprehensive temporal monitoring system using remote sensing data to track changes in impervious surfaces and vegetated areas within urban agglomerations over time. This objective aims to unravel the dynamics of urban development and land-use changes, providing a basis for assessing the effectiveness of urban planning strategies in response to climate change.

1.2.3 Integration of climate change impacts into urban planning

Investigate the integration of GIS and remote sensing data into urban planning processes to enhance climate change adaptation strategies. This includes assessing the vulnerability of urban structures to climate change impacts, such as increased temperatures and extreme weather events. Develop recommendations for incorporating spatial information into urban planning frameworks to foster resilient and sustainable urban development.

1.2.4 Intra-urban thermal environment analysis

Explore the spatial variability of the links between urban vulnerability and characteristics within municipal boundaries. Utilize GIS techniques to assess the intra-urban thermal environment and identify hotspots of vulnerability. This research objective aims to provide a nuanced understanding of how different urban zones are affected by climate change, guiding the development of targeted adaptation measures for specific areas within the urban agglomeration.

1.3 Research scope

This research seeks to comprehensively assess approaches for the impacts of climate change on urban agglomerations by employing GIS and remote sensing techniques. **The first objective** delves into a focused spatial analysis, identifying and characterizing urban heat islands (UHIs) within different land-cover types. Emphasis is placed on the relationship between land surface temperature (LST) variations and various urban structures. **The second objective** concentrates on developing a streamlined temporal monitoring system, utilizing remote sensing data to track changes in impervious surfaces and vegetated areas over time. This provides insights into the dynamic nature of urban development and land-use changes. **The third objective** explores the practical integration of GIS and remote sensing data into urban planning processes, enhancing climate change adaptation strategies. This involves vulnerability assessments and recommendations for incorporating spatial information into planning frameworks. Lastly, the **fourth objective** focuses on the spatial variability of links between urban vulnerability and characteristics within municipal boundaries, utilizing GIS techniques to assess the intra-urban thermal environment and identify vulnerable hotspots. The overarching goal is to guide the development of targeted adaptation measures for sustainable urban development.

2. Methodology

This methodology presents a generic methodology that provides an integrated approach utilizing remote sensing and GIS techniques to comprehensively assess the impacts of climate change on urban agglomerations. The research initiates with the acquisition of satellite images for identifying land transformations and monitoring changes in the Earth's surface, with a specific focus on urbanization and land use/land-cover changes. The leverage of remote sensing data, particularly satellite images, stems from their extensive spatial coverage and regular availability, enabling effective monitoring of urban structures and climate indicators. GIS is subsequently employed to analyze and visualize spatial data extracted from remote sensing, facilitating the assessment of climate-related indicators in urban areas. The synergy of remote sensing and GIS allows for the extraction of intricate information from urban surfaces, including land surface temperature (LST), imperviousness, and vegetation vitality indicators. To enhance efficiency, advanced technologies such as Artificial Intelligence (AI) are integrated to automate the measurement and evaluation of climate adaptation indicators in urban areas. Climate change vulnerability and adaptation assessments are conducted through a fusion of remote sensing data, GIS analysis, and numerical modeling systems, allowing for an in-depth analysis of the urban climate and thermal bioclimatic in urban agglomerations. The incorporation of Local Climate Zones (LCZ) based on remote sensing data and GIS techniques facilitates the analysis of climate change impacts on LST under different land-cover types in urban agglomerations. This comprehensive methodology aspires to provide a robust framework for understanding and addressing the intricate dynamics of climate change in urban environments.

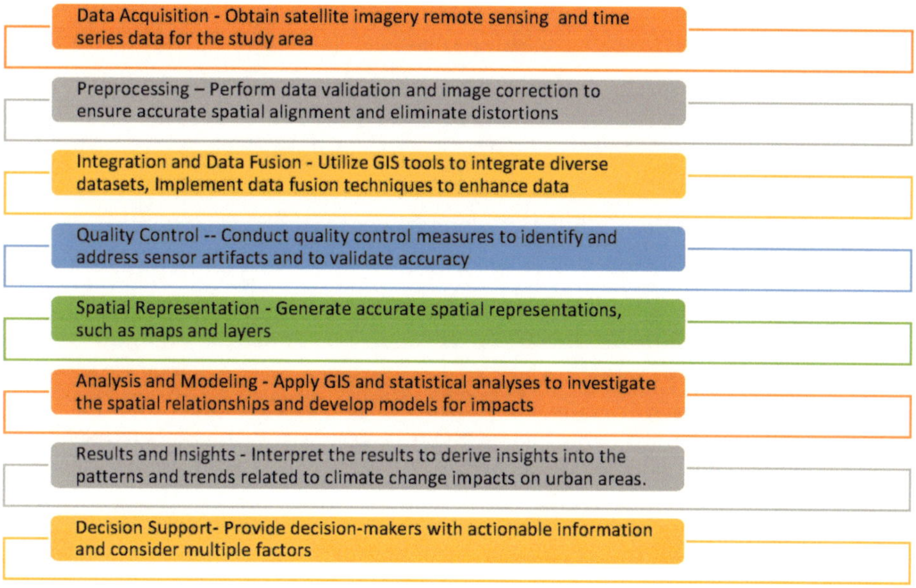

Figure 1.
Methodology flowchart.

The research methodology entails systematic data collection and preprocessing to ensure the reliability and accuracy of the analysis. Satellite imagery and remote sensing data specific to the study area will be acquired, capturing pertinent information on land-cover, surface temperature, and urban structures. To introduce a temporal dimension, time-series data will be collected to monitor changes over defined periods. The collected data will undergo rigorous preprocessing, incorporating radiometric and atmospheric corrections, geometric adjustments, and image normalization. GIS tools will be employed to integrate diverse datasets, facilitating the creation of precise spatial representations. Quality control measures will address sensor artifacts, enhancing the overall reliability of the dataset. This comprehensive approach to data collection and preprocessing is pivotal for generating robust insights into the impacts of climate change on urban agglomerations, ensuring the validity of subsequent analyses and supporting informed decision-making in urban planning and climate change adaptation strategies (**Figure 1**).

3. Discussion

This study attempts to unravel the complexities of climate change impacts on Urban Agglomerations. It employed a comprehensive methodology, integrating GIS and remote sensing techniques, to assess the multifaceted impacts of climate change on urban agglomerations [10]. The spatial analysis of urban heat islands (UHIs) unveiled intricate relationships between land surface temperature (LST) variations and different land-cover types, emphasizing the significance of understanding nuances within urban structures [11]. The temporal monitoring of urban structures uncovered dynamic patterns of change in impervious surfaces and vegetated areas, underscoring the evolving nature of urban development in response to climate influences. Integrating climate change impacts into urban planning processes proved to be

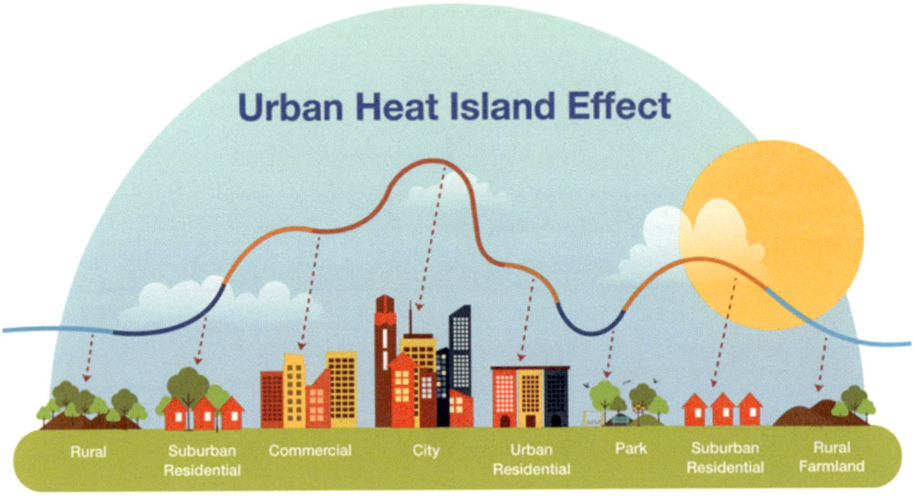

Figure 2.
Urban Head Islands effect, as it relates to urban agglomerations (after the City of Little Rock, Arkansas). https://www.littlerock.gov/city-administration/city-departments/public-works/sustainability/urban-heat-islands/

a pivotal aspect, emphasizing the need for vulnerability assessments and the incorporation of spatial information for effective adaptation strategies [12]. The intra-urban thermal environment analysis further nuanced our understanding, highlighting specific vulnerable hotspots within municipal boundaries [13]. The success of the methodology relied heavily on robust data collection and preprocessing, ensuring the reliability of the analyses conducted (**Figure 2**) [14].

3.1 Preprocessing steps and data quality

The preprocessing steps, including radiometric and atmospheric corrections, geometric adjustments, and data fusion, played a critical role in enhancing the accuracy of the dataset. Quality control measures addressed potential artifacts, contributing to the overall credibility of the findings. By systematically addressing sensor-specific variations and atmospheric effects, radiometric corrections ensured that the data accurately reflected the true surface characteristics. Geometric corrections eliminated distortions, guaranteeing accurate spatial alignment crucial for meaningful spatial analyses. Data fusion techniques enriched the integrated dataset, providing a comprehensive view of urban structures and climate indicators. The meticulous approach to quality control reinforced the reliability of the dataset, enhancing the robustness of subsequent analyses.

3.2 Spatial representation and analysis

The spatial representation of integrated datasets facilitated a visually compelling depiction of the complex interactions within urban areas. Accurate spatial representations, including maps and layers, served as effective tools for conveying the nuanced relationships between urban structures, land-cover, and surface temperature. These visualizations were instrumental in communicating complex findings to diverse stakeholders, aiding in the interpretation of spatial patterns and trends. The analysis and modeling phase provided quantitative insights into spatial relationships, enabling the development of models that can inform future projections of climate change impacts on urban landscapes.

3.3 Decision support and implications

The results presented in this study contribute valuable information for decision-makers involved in urban planning and climate change adaptation. By unraveling the complexities of climate change impacts on urban agglomerations, this research provides a foundation for evidence-based decision-making. The findings emphasize the importance of spatially informed strategies for sustainable urban development, underscoring the need for targeted interventions in vulnerable areas. As we navigate an era of rapid urbanization and climate change, the integration of GIS and remote sensing techniques proves indispensable for shaping resilient and adaptive cities.

The study offers crucial managerial implications for decision-makers involved in urban planning and climate change adaptation. By unraveling the complexities of climate change impacts on urban areas, the research provides a foundation for evidence-based decision-making. Decision-makers can utilize the insights gained from GIS and remote sensing techniques to inform urban planning strategies effectively. The identification of vulnerable areas, conducted through spatial analysis, allows for targeted interventions, optimizing resource allocation for climate change adaptation. The study's emphasis on spatially informed strategies underscores the need for decision-makers to integrate GIS and remote sensing data into planning frameworks. This integration ensures that climate change considerations are seamlessly embedded in urban development plans, promoting sustainable and resilient cities. The spatial analysis further aids in the identification of vulnerable hotspots within urban agglomerations, guiding decision-makers in prioritizing areas requiring immediate attention. Resource allocation becomes more effective as decision-makers gain an understanding of the spatial variability of climate change impacts, allowing for nuanced, tailored approaches to diverse urban zones. The study advocates for a proactive stance by decision-makers, leveraging real-time monitoring capabilities provided by GIS and remote sensing for swift responses to emerging challenges. Additionally, the findings stress the importance of community engagement and awareness, enabling decision-makers to foster collaboration and inclusivity in climate resilience initiatives. In essence, the research offers a comprehensive framework for decision-makers to navigate the complexities of urbanization and climate change, emphasizing the indispensable role of GIS and remote sensing in shaping resilient and adaptive cities for the future.

4. Conclusion

In conclusion, this study showcases the significance of employing GIS and remote sensing techniques to comprehensively understand and address the impacts of climate change on urban agglomerations. Moving forward, continuous advancements in technology, including the integration of Artificial Intelligence, offer opportunities to further enhance the precision and efficiency of climate change impact assessments. Future research endeavors should explore the incorporation of additional socio-economic and demographic factors to deepen our understanding of vulnerability within urban populations. Additionally, longitudinal studies could provide insights into the long-term effectiveness of implemented urban planning strategies. Overall, as urban areas continue to evolve in response to climate challenges, the integration of spatial technologies remains pivotal for informed decision-making and sustainable urban development.

5. Future directions

5.1 Adaptation strategy for climate-resilient urban development

In response to the identified vulnerabilities revealed through our GIS and remote sensing analysis, an adaptive strategy is proposed to enhance the resilience of urban agglomerations. This strategy encompasses targeted interventions addressing specific vulnerabilities, while also integrating broader measures related to urban heat island (UHI) mitigation, flood resilience, and sustainable urban planning.

5.1.1 Tailored vulnerability interventions

- **Green Infrastructure Implementation:** Identify and prioritize vulnerable hotspots within the urban fabric, particularly areas experiencing elevated land surface temperatures. Implement green infrastructure initiatives, including the creation of urban green spaces and increased tree canopy coverage, to mitigate heat island effects and enhance overall urban resilience.

- **Localized Flood Management:** Develop localized flood management strategies tailored to areas prone to flooding. This may include the implementation of permeable surfaces, green roofs, and strategically designed drainage systems to reduce surface runoff and enhance flood resilience.

5.1.2 Urban heat island mitigation

- *Cool Roof Initiatives:* Encourage and incentivize the adoption of cool roofing technologies, especially in areas with high impervious surface concentrations. Cool roofs reflect more sunlight and absorb less heat, contributing to a reduction in surface temperatures and mitigating UHI effects.

- *Tree Canopy Expansion:* Prioritize the expansion of urban tree canopy coverage, focusing on areas identified as UHI hotspots. Trees provide shade, enhance evapotranspiration, and contribute to a cooler microclimate, effectively countering the urban heat island effect.

5.1.3 Flood resilience measures

- *Elevated Infrastructure Design*: In flood-prone areas, promote the construction of elevated or flood-resistant infrastructure. Elevating critical facilities and residential structures above flood levels reduces the impact of inundation and safeguards against potential damages.

- *Community Early Warning Systems:* Implement community-based early warning systems to enhance preparedness for flood events. Disseminate timely information and evacuation protocols, ensuring that vulnerable populations are well-informed and can respond effectively to flood threats.

5.1.4 Sustainable urban planning

- *Compact and Mixed-Use Development:* Encourage compact and mixed-use urban development patterns to minimize the expansion of impervious surfaces. This approach promotes walkability, reduces the urban heat island effect, and enhances the overall sustainability of urban areas.

- *Integrated Land-Use Planning:* Integrate climate resilience considerations into land-use planning processes. Ensure that future urban development adheres to sustainable practices, minimizing environmental impact and promoting long-term resilience to climate change.

By implementing this adaptive strategy, we aim to create climate-resilient urban agglomerations that not only address specific vulnerabilities but also foster sustainable development practices. This multifaceted approach acknowledges the interconnected nature of climate change impacts, positioning urban areas for a more resilient and sustainable future. The success of this strategy hinges on collaborative efforts among policymakers, urban planners, and communities to enact and sustain these adaptive measures.

5.1.5 Forging resilient urban futures in the face of climate change

This research, employing a holistic approach integrating GIS and remote sensing techniques, has illuminated the intricate dynamics of climate change impacts on urban agglomerations. The spatial analysis uncovered the complex relationships between land surface temperature variations and diverse land-cover types, emphasizing the need for a nuanced understanding of urban structures. Temporal monitoring revealed the dynamic nature of urban development, urging the adoption of adaptive strategies to navigate evolving climate challenges. The integration of climate change impacts into urban planning processes underscored the necessity of vulnerability assessments and spatially informed strategies for resilient urban futures.

In addressing specific vulnerabilities, the proposed adaptive strategy encapsulates tailored interventions while aligning with broader goals of urban heat island mitigation, flood resilience, and sustainable urban planning. Initiatives such as green infrastructure implementation, cool roof adoption, and elevated infrastructure design aim to fortify urban areas against the impacts of a changing climate. These strategies not only mitigate specific vulnerabilities but also contribute to the overarching goal of fostering sustainable and climate-resilient urban development.

Crucially, the success of these adaptation measures hinges on robust data collection and preprocessing, ensuring the reliability of our analyses. Quality-controlled, integrated datasets form the foundation upon which informed decision-making can occur. The spatial representation of these datasets serves as a visual roadmap for identifying vulnerabilities and formulating targeted interventions.

As urbanization and climate change continue to shape our world, the findings of this research contribute substantively to the discourse on resilient urban futures. The proposed adaptive strategy provides a blueprint for cities globally, offering a flexible framework that can be tailored to diverse urban contexts. By acknowledging the interconnected nature of climate change impacts and embracing adaptive strategies, urban areas can forge resilient paths forward, fostering sustainable development in the face of an ever-changing climate. As we navigate the challenges of the twenty-first century, the fusion of advanced technologies, data-driven insights, and strategic planning emerges as a beacon guiding cities toward climate resilience and a sustainable urban future.

5.2 Future work: charting paths for continued research and action

While this research has provided valuable insights into the impacts of climate change on urban agglomerations and proposed an adaptive strategy, there remain avenues for future exploration and enhancement:

5.2.1 Fine-scale analysis

Future research could delve deeper into fine-scale analyses, examining micro-level variations in land surface temperature and vulnerabilities within urban areas. This could involve higher-resolution remote sensing data and more advanced GIS techniques to capture localized nuances.

5.2.2 Dynamic modeling

Developing dynamic models that incorporate predictive capabilities would enhance our ability to foresee the evolving impacts of climate change on urban landscapes. Incorporating climate change projections and urban development scenarios would offer a forward-looking perspective for more effective long-term planning.

5.2.3 Community engagement and social dynamics

Expanding the research scope to include community engagement and social dynamics would provide a more comprehensive understanding of resilience. Investigating how communities perceive and respond to climate change impacts can inform adaptive strategies that align with local needs and priorities.

5.2.4 Multi-criteria decision analysis

Integrate multi-criteria decision analysis (MCDA) approaches to prioritize and optimize adaptation measures. This could involve considering not only climate-related factors but also economic, social, and governance dimensions to ensure comprehensive and inclusive decision-making.

5.2.5 Technological advances

Embrace emerging technologies, such as advanced satellite sensors and machine learning algorithms, to enhance the accuracy and efficiency of data collection, analysis, and modeling. This could provide more detailed and real-time information for decision-makers.

5.2.6 Long-term monitoring

Establishing long-term monitoring programs to track the effectiveness of implemented adaptation measures is crucial. This would provide continuous feedback on the success of strategies, allowing for adaptive management in response to changing conditions.

5.2.7 Cross-disciplinary collaboration

Foster collaboration between climate scientists, urban planners, social scientists, and policymakers to create an integrated and holistic approach. Cross-disciplinary efforts can provide a more comprehensive understanding of the challenges and facilitate the development of effective solutions.

5.2.8 Global comparative studies

Undertake global comparative studies to assess the transferability and scalability of adaptation strategies across diverse urban contexts. Understanding how different cities respond to similar challenges can provide valuable insights for shared learning and best practices.

Author details

Rifaat Abdalla
Department of Earth Sciences, College of Science, Sultan Qaboos University, Al-Khoudh, Oman

*Address all correspondence to: rabdalla@squ.edu.om

References

[1] Boulanger SOM. Urban adaptation to climate change state of the art: Evaluating the role of adaptation assessment frameworks through a systematic and bibliometric analysis. Sustainability. 2023;**15**(13):10134. DOI: 10.3390/su151310134

[2] Barat A, Kumar S, Aakriti Asim P, Sarthi P. Monitoring of urban climate using geospatial techniques. In: Advances in Urbanism, Smart Cities, and Sustainability. Boca Raton, Florida, USA: CRC Press; 2022. DOI: 10.1201/9781003126195-30

[3] Lu-Yun L, Yu S, Komi H, Bedra B, Zhang M. Investigating the spatial heterogeneity of urban heat island responses to climate change based on local climate zones. Sustainability. 2023;**15**(7):6298. DOI: 10.3390/su15076298

[4] Ricardo V. Remote sensing and AI for building climate adaptation applications. Results in Engineering. 2022;**15**:100524-100524. DOI: 10.1016/j.rineng.2022.100524

[5] David H, Julián G, Díaz A, Martín A, Emilio M, Cobos G. Spatiotemporal analysis of urban thermal effects caused by heat waves through remote sensing. Sustainability. 2022;**14**(19):12262. DOI: 10.3390/su141912262

[6] Linda O, Mearns Stephan R, Sain Lai-Yung R, Melissa L, Seth B, Sébastien MG, et al. Climate change projections of the North American regional climate change assessment program (NARCCAP). Climatic Change. 2013;**120**(4):965-975. DOI: 10.1007/S10584-013-0831-3

[7] Stenka V, Alby D, Meier RF, Nouri H, Schulz C, Soulsby C, et al. City-wide, high-resolution mapping of evapotranspiration to guide climate-resilient planning. Remote Sensing of Environment. 2023;**287**:113487. DOI: 10.1016/j.rse.2023.113487

[8] Ritu G. Evaluating the impact of climate change on the urban environment using geospatial technologies in Bhubaneswar, India. The International Archives of the Photogrammetry, Remote Sensing and Spatial Information Sciences. 2022;**XLVIII-4/W5-2022**:159-166. DOI: 10.5194/isprs-archives-xlviii-4-w5-2022-159-2022

[9] Abida F, Ul M, Harmeet S, Ahmed SP. Assessment of spatiotemporal changes in land use/land cover of North Kashmir Himalayas from 1992 to 2018. Modeling Earth Systems and Environment. 2020;**6**(2):1189-1200. DOI: 10.1007/S40808-020-00750-9

[10] Dong G, Xie Y, Wang Y, Fan D, Tian Z. Ensemble projection of extreme precipitation over China based on three dynamical downscaling simulations. Frontiers in Earth Science. 2021;**9**:823. DOI: 10.3389/FEART.2021.755041

[11] Ghasem F, Shahriar KM, Manijeh GT, Patel ND. Effects of climate change on dynamics of agricultural lands and cultivation pattern, a case study of Urmia County, Iran. Arabian Journal of Geosciences. 2022;**15**(21):1643. DOI: 10.1007/s12517-022-10926-5

[12] Thomas E, Thomas E, Vitus H, Gunther S, Michael T, Thilo W, et al. Large-area assessment of impervious surface based on integrated analysis of single-date Landsat-7 images and geospatial vector data. Remote Sensing of Environment. 2009;**113**(8):1678-1690. DOI: 10.1016/J.RSE.2009.03.012

[13] Daisy S, Martin S, Aguayo LG, Wallace L, Reinke K, McLennan B. Perceptions of land use and land cover analysed using geospatial data. Applied Geography. 2022;**146**:102757-102757. DOI: 10.1016/j.apgeog.2022.102757

[14] Lin TH, Liu GR, Chen YC. Remote sensing of smoke plumes with moderate resolution imaging spectroradiometer reflectance measurements. Journal of Applied Remote Sensing. 2010;**4**(1): 041876-041876. DOI: 10.1117/1.3505481

Chapter 3

Sustainable Urbanization between Two Ambitious Global Agendas: An Integration Approach

Abdulkarim Hasan Rashed

Abstract

The 2030 Agenda for Sustainable Development and the New Urban Agenda (NUA) form significant ambitions towards a sustainable and better human future. The sustainable development goals (SDGs) have a deep-rooted urban dimension, as exemplified in Goal 11, which intends to make cities inclusive, safe, resilient, and sustainable, and around 23% of the SDG indicators have a clear urban component. That is forming a window of opportunity to integrate both Agendas towards sustainable urbanization and improving the quality of life in urban agglomerations. Both Agendas should be functioning jointly towards contributing to improving the quality of life and providing all essential life services and needs. Thus, the integration approach is the best pathway to synergy aims of both Agendas to tackle the challenges and formulate a coherent interlinkage in the thematic areas. Consequently, urbanization and sustainable urban agglomeration development need a strong Agenda and implementation framework to tackle the environmental, social, and economic challenges of urban agglomerations. Therefore, the big lesson distillery from both Agendas is to localize the SDGs in the urban agglomerations to attain sustainability at a broad scope. Overall, the integrated approach to urban sustainability covers all sustainability aspects and the correlations among sustainability dimensions in both Agendas.

Keywords: urban sustainability, sustainability planning, SDGs localization, urban agglomerations development, coherent policy, urban resilience

1. Introduction

In 1987, the Brundtland Report recognized the significance of cities in achieving sustainability, followed by the Rio Declaration on Environment and Development in 1992, which declared the urgent necessity for enhancing urban sustainability [1, 2]. In 2000, the Millennium Development Goals (MDGs) were employed as tools to improve better services for the poor urban; however, the urban and city dimensions were addressed only in MDG 7 (target 11) is articulated that "ensure environmental sustainability: achieving by 2020 a significant improvement in the lives of at least 100 million slum dwellers" [3]. The first World

Cities Day (WCD) was held on October 31, 2014, under the theme "Leading Urban Transformations", to date, urbanization issues have remained a top global priority for achieving sustainable city transformations. Since then, urban sustainability has become a priority concern for scientists, decision-makers, and policy-makers, as stressed in policy frameworks and international reports [4]. In 2023, the WCD theme is "Financing Sustainable Urban Future for All", according to the United Nations Conference on Trade and Development (UNCTAD) and the International Energy Agency (IEA), approximately $2.6 trillion per year is needed until 2030 to achieve the Sustainable Development Goals (SDGs) and reach a net-zero society by 2050. Meanwhile, the 2023 report of "United Nations World Economic Situation and Prospects" predicts that the global growth rate will slow down to 1.9% in 2023, nevertheless, the World Bank reports sufficient capital available to address global infrastructure needs [5–7].

Urbanization is a transformative process that drives global development. With increasing concern from global agencies in urbanization and sustainable development, there is a growing urgency to thoroughly assess how sustainable urbanization can be achieved worldwide [8]. Further, there is ample evidence of cities in UN affairs, frameworks, and programs. For instance, the localization of the "Sendai Framework for Disaster Risk Reduction", the Paris Agreement on Climate Change, the Protection of the World Cultural and Natural Heritage Convention, and the WHO "Shanghai Consensus on Healthy Cities" [9]. Thus, it is crucial to integrate the New Urban Agenda (NUA) with the 2030 Agenda for Sustainable Development to effectively tackle the major challenges—e.g., environmental sustainability, waste management, climate change adaptations, and access to services—in urban environments. These challenges present significant opportunities to implement sustainable practices, such as green or smart city principles, which can lead to a better quality of life. Therefore, there is a need to build an effective nexus between the two Agendas through a participatory approach among multi-sectoral and multi-stakeholders in the urban planning process and policy-making. In this vein, the UN has taken a significant step by incorporating urban issues into the pivotal 2030 Agenda, particularly in SDG 11 ("Sustainable Cities and Communities"), which establishes a connection with the NUA. Therefore, this goal creates a significant bridge between the two Agendas towards their integration approach.

Urban areas are home to 55% of the world's population (**Figure 1**) and generate 85% of the global gross domestic product (GDP) as they serve as centres of employment, innovation, and investments. In contrast, the mushrooming literature evident that urbanization has a negative impact on sustainable development such as blundered urban development, demolition of ecologically sensitive zones, and cities account for 75% of the total greenhouse gas emissions, and produce 70% of global waste due to growing industrial activities and unsustainable consumption, leading to increased environmental degradation (**Figure 2**) [11–16]. According to the United Nations 2050 global prediction, 75% of the global population will live in cities, where 18% will occupy the 40 largest urban agglomerations, 66% economic business, and 85% technological innovations [17]. Additional impacts of rapid urban expansion are excessive energy and water consumption, weak infrastructure, lack of services, traffic congestion, and climate change [18]. Therefore, the city provides significant opportunities to address sustainability concerns across various sectors—e.g., energy, transportation, and water—via legal, financial, planning, and administrative instruments. Thus, addressing urban sustainability is the key to tackling global sustainability.

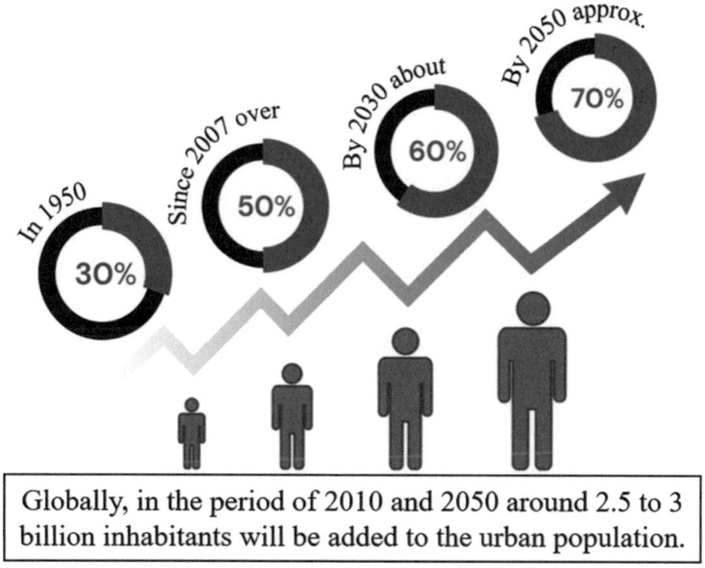

Figure 1.
The growth and outlook of global urbanization in the period from 1950 to 2050 (data source: [10]).

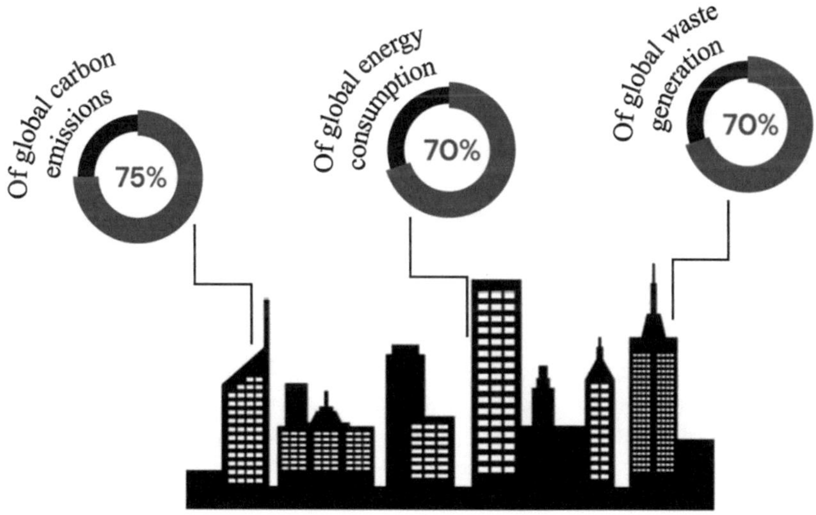

Figure 2.
The cities' impact on environment.

In that sense, Agenda 2030 and the NUA represent a historical opportunity and mark value approach to achieving sustainable development.

The rest of the chapter is organized as follows. The second section discusses the sustainable dimension of the New Urban Agenda. The third section discusses SDG 11 as an urban sustainability framework. The fourth section discusses the pathway towards the integration approach. While, the fifth section reviews the best urban sustainability practices, and the sixth section reviews the sustainable urbanization lessons learnt. The last section concludes the chapter.

2. The sustainable dimension of the new urban agenda

The NUA is a holistic urban view, which is considered a benchmark to attain sustainability efficiently through the realization of its social task. Whereas, cities are considered drivers for sustainable development [19]. The core driving principles of the NUA are similar to the 2030 Agenda; for example, both Agendas are universal, voluntary, and led by the state, and motivate member states to tackle the global common challenges [20]. According to Ali-Toudert et al., urban sustainability can be defined as "a design of future urban development as well as the re-development of existing ones in an environmentally friendly and resource-efficient manner" [21]. The United Nations defines urban agglomeration as "the de facto population contained within the contours of a contiguous territory inhabited at urban density levels without regard to administrative boundaries" [22]. Urban agglomeration benefits from integrating industrial distribution, resilient infrastructures, establishing regional markets, planning urban and rural, protecting the environment and ecosystems, and social and security systems; where, in the future, it will be the primary carrier for socio-economic development [17, 23].

The NUA is a significant landmark document that addresses the major challenges cities face; it should be used to measure the accomplishment progress of future cities; its success depends on the formation of comprehensive and lasting synergies [24]. The NUA was adopted on 20 October 2016, and it was the first globally acquiesced document describing the implementation of the urban dimension of the SDGs. Item 9 articulates that "*the New Urban Agenda reaffirms our global commitment to sustainable urban development as a critical step for realizing sustainable development in an integrated and coordinated manner at the global, regional, national, subnational and local levels, with the participation of all relevant actors. The implementation of the New Urban Agenda contributes to the implementation and localization of the 2030 Agenda for Sustainable Development in an integrated manner, and to the achievement of the Sustainable Development Goals and targets, including Goal 11 of making cities and human settlements inclusive, safe, resilient and sustainable*" [25].

According to the UN-Habitat, revisiting the NUA is embracing the urbanization policies that can assist countries in handling challenges by developing national frameworks, enhancing national urban planning, selecting the appropriate SDGs that support sustainable urbanization, and aligning and strengthening institutional structures. As a consequence, the implications of implementing the NUA will reflect on urban rules and regulations, urban planning and design, and municipal finance [25, 26]. **Figure 3** illustrates the core aims of NUA.

Three decades of sustainable development have passed, and the controversy has acquired a global dimension about the balance among political cultures, scientific, and experts' knowledge in environmental management and other menaces issues, where the responsibilities became more complex and need for robust evidence-based scientific justifications; this challenge is at the core of the NUA [27]. Consequently, there is a need for tools to implement NUA to ensure the shifts in urban development and undertake the challenges of modern urbanization. There are two tools: the "International Guidelines on Urban and Territorial Planning" (IG-UTP)—a set of policy recommendations to aid countries and concerned stakeholders in urbanization and human settlements—and "National Urban Policies" (NUP)—a coordinated and deliberate government-led process to rally various stakeholders towards resilient urban development for the long term—[28, 29], where the NUP is considered a significant strategic tool for advancing sustainable urbanization and monitoring SDG11

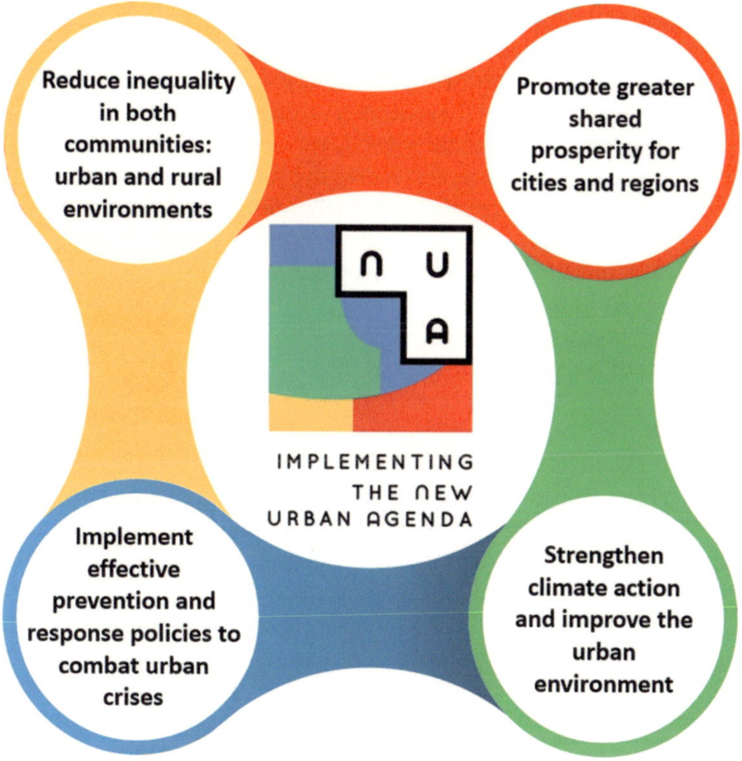

Figure 3.
The core aims of new urban agenda.

implementation [26]. Therefore, sustainable urbanization in the context of NUA is a transformative force to attain the SDGs of the 2030 Agenda.

To a large extent, urbanization has become a key driver of sustainable development and global environmental change. Therefore, the focus on cities in the new global development goes far beyond the usual focus on providing housing and improving slums; the vision expands to include safe and sustainable transportation, integrated and participatory urban planning, green public spaces, air quality, water management, waste management, climate resilient, and natural disaster risk. This vision will assist in improving current urban processes, policy, and planning globally [3]. For instance, NUA is committed in its vision statement to "Protect, conserve, restore and promote their ecosystems, water, natural habitats and biodiversity, minimize their environmental impact and change to sustainable consumption and production patterns" [25], and in 2021, the Global Platform for Sustainable Cities (GPSC) was launched a program (Cities-4-Biodiversity) to provide technical and financial support to encourage cities to integrate their urban planning biodiversity and climate change solutions [30].

3. SDG 11: the urban sustainability framework

The 2030 Agenda—and its SDGs—are closely aligned with the NUA, and both Agendas are comprehensive, ambitious, and socially progressive Agendas that have the prospect of contributing towards cities' transition to be sustainable, inclusive, and resilient [31, 32]. Therefore, the national urban policy can act as a qualitative toolbox for guiding and monitoring the attainment of the 2030 Agenda [12]. The Paris Agreement on Climate Change, the Sendai Framework on Disaster Risk Reduction, and the NUA recognize the significant role of urban resilience in the face of climate change and natural disasters; where the resilience concept cuts across the 17 SDGs. Further, urban areas are critical for achieving the SDGs as more than 50% of targets are related to urban areas. In contrast, both SDGs 11 ("Sustainable Cities and Communities") and 17 ("Partnerships for the Goals") consider around 42% (73 items) of the NUA [33]. Therefore, a comprehensive sustainable framework is essential to building urban resilience [34, 35].

The assignment of a separate urban goal (SDG 11) within the 2030 Agenda, alongside the NUA, reflects the success of advocacy efforts to raise policy focus and financial support for cities, recognizing their crucial role in facilitating sustainable development [3]. SDG 11 is purely dedicated to sustainable cities, which is a framework for sustainable urban development that addresses a variety of interconnected issues. It takes a systems approach, emphasizing cross-linkages with other developmental priorities. Its goal is to achieve sustainable urban development by 2030 through a set of achievable targets and indicators functioned to steer policy actions at the local level. In addition, SDG11 and its targets are opportunities to focus on effective indicators, data, measurements, and standards that assist in understanding cities particularly, and global urbanization at large, which could be addressed through an integrated approach [36]. Further, the SDG11 along sisters' goals, aims to make cities safe, inclusive, resilient, and sustainable, placing urbanization at the forefront of international development policy. In this vein, the inclusion of the targets of SDG 11 into the NUA is a difficult mission due to challenges that lack data and measuring urban realities development, where measurement becomes crisscrossed with human lives and priorities [36]. Thus, it is important to contextualize SDG 11 within the local urban development process.

To achieve SDG 11, UN-Habitat, jointly with other custodian international agencies such as UNISDR, UNESCO, WHO, UNODC, UNEP, and concerned stakeholders, has supported several methodological activities aimed at establishing systems for gathering qualitative and quantitative data. These activities include the development of guides for using geospatial information technology (GIS), big data analytics, and community-based data as additional sources of data at the national level. Further, UN-Habitat has also directly worked with countries to implement these systems for data collection [10]. Achieving SDG 11 makes cities not only sustainable but also contributes to other goals [37]. Therefore, cities must be a priority at the heart of efforts to accomplish global sustainability [38]. **Figure 4** illustrates the linkages between SDG 11 targets with other goals.

Figure 4 illustrates that SDG11 and its targets are interlinked with the entire SDGs. For instance, SDG 1 (targets: 1.2, 1.3, and 1.4) addressed reducing people living in poverty in cities, assuring access to basic services and assured social protection, while target (1.b) concerning the importance of improving the poor urban areas by formulating development strategies and policies. SDG 2 (target: 2.1) assures access to sufficient food for vulnerable people. Whereas SDG 3 (targets: 3.6, 3.7, and 3.9) focused on sexual and reproductive health and death prevention as a result of traffic accidents and pollution, those issues are considered increasing concerns in cities. While, SDG 4 (targets: 4.3 and 4.6) addressed eradicating literacy, and facilitating access to technical and higher training, especially for the most vulnerable groups,

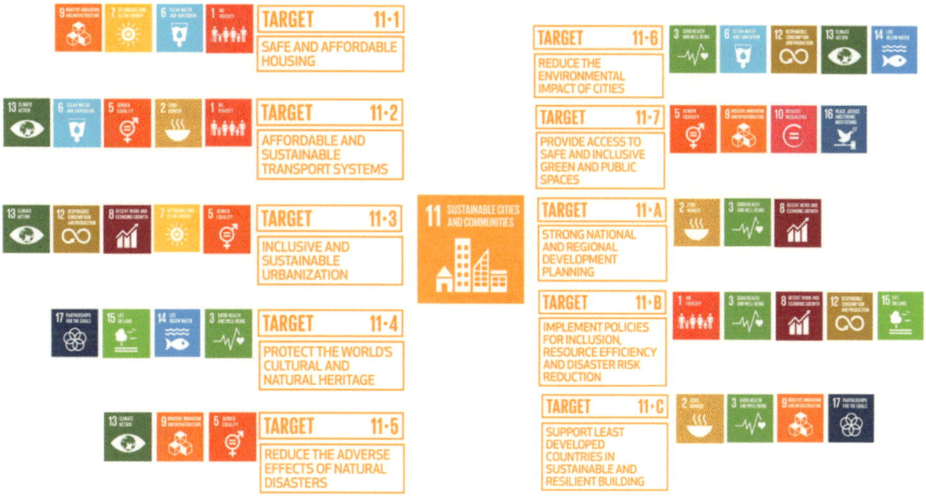

Figure 4.
The linkages between SDG 11 targets with other sustainable goals.

which are key challenges in the cities. SDG 5 (targets: 5.2, 5.4, 5.5, and 5.6) is considered a typical goal of eradicating all forms of violence against women, respecting care work, ensuring the highest women's participation rights, and ensuring health services as articulated in SDG 3. Both SDGs 6 (targets: 6.3 and 6.b) and 7 (target: 7.1) focused on essential resources -water and energy- to sustain the cities; those goals aimed at improving wastewater treatment and providing affordable energy, also both are related to SDG 12 regarding sustainable consumption. SDG 8 (targets: 8.3, 8.5, 8.8, and 8.b) discoursed important economic parts of cities through employment promotion, ensuring safe workplace environments, and strategies for the employment of young people. SDG 9 addressed technological innovation through this goal can build a smart city. SDG 10 is about reducing inequality, and this goal is reflected in all SDGs especially SDG 5. SDG 12 (target: 12.5) focused on waste production in the city and called for reduction and recycling. With regard to SDG 16 (targets: 16.1, 16.2, 16.3, 16.9, and 16.10) covered important matters such as violence and abuse prevention, justice, social inclusion, information, citizen participation, and legal support, and this goal is supportive of other SDGs.

In sum, SDG 11 is viewed as an urban sustainability framework, which presents windows of opportunity for overcoming urban challenges and enhancing synergies and integration capacities—e.g., policies, urban planning, resilient infrastructures, and services—and a hub between the two Agendas towards a tangible sustainable transition.

4. The pathway towards the integration approach

The issues associated with urban growth and sustainability should be treated together as one issue. While, two significant challenges may hinder the achievement of the two Agendas by the 2030 stipulated timeframe which are (a) the current urban development paths are described as inactive planning systems and affected by personal benefits, and (b) present global economic systems often discrepancy with achieving high sustainability standards. In addition, due to urban development complexity, profound uncertainty, and scale, radically unique approaches are needed to attain planet sustainability. Nevertheless, taking the two Agendas seriously and continuing the implementation can provide a valuable opportunity to rethink urban planning and development considering the sustainability dimensions (social, environmental, and economic). This opportunity will serve to reassess governance systems and prioritize sustainability in urban planning and development Agendas [26, 32, 39–41]. To accomplish sustainable urban development, one needs to create a holistic understanding of sustainability through a sound comprehension of notions, approaches, strategies, tools, and techniques employed to assess urban development's sustainability [42], which will result in integrated and cohesive solutions and harmonized actions in addressing urban sustainability.

Urban development is crucial for sustainability due to the significant impacts of high population density on society and the economy [33]. In addition, complex and multidimensional disadvantages in urban areas require a comprehensive approach to address such challenges [43]. While the current progress in implementing the SDGs in urban areas made so far is trim there is a need to develop integrated approaches [44]. To understand urban scaling in the transformation to sustainability. Two links should be understood, development and sustainability, as well as, development and urbanization raise a significant question of whether urbanization is useful for sustainability or antagonizing it [37, 45]. The linkage between urbanization and development represents the nexus between the NUA and the 2030 Agenda [46]. The 2030 Agenda combines human development (embodied in city and population growth) and sustainability transformation (embodied in human welfare, economic prosperity, and environmental conservation) [11]. Therefore, both Agendas are "top and foundation stones to adopt practical measures, coherent policies, and effective legal frameworks to improve resilient cities, better human settlements, aspirations life, and resilient infrastructure through sustainable planning, management, and governance to mitigate or prevent the devastating impacts of disasters and pandemics" [47].

Unsustainable cities can be avoided in the coming decades by investing in infrastructures on a green path and employing new efficient approaches, such as the urban nexus and urban ecosystems to create urban areas that are habitable, healthy, sustainable, and prosperous [20]. To tackle the challenges of cities' change and restructuring, it should embrace a comprehensive and coordinated approach to urban development aiming to concentrate on long-term strategies and enhance sustainability aspects -social, economic, and environmental- towards a better sustainable future [48]. Both Agendas reiterate the global commitment to sustainable urban development as a key driver of sustainable development at global, regional, and national levels, where all concerned stakeholders should engage to actively participate in the implementation process. Some key commitments included providing essential life needs and services (e.g., sanitation, health, electricity, water supply, and education), strengthening urban resilience, implementing mitigation measures to tackle environmental impacts (e.g., climate change and pollution), establishing sustainable infrastructures, formulating coherent policies, and enhancing best governance practices.

The NUA cannot be effectively implemented standalone to attain tangible sustainable urban development. Because NUA focuses most of its actions on the managerial aspect, which can synergize with the SDGs. This framework (**Figure 5**) emphasizes the integration of the two Agendas and the consideration of resilient urban development aimed at contributing to developing a sustainable urban environment. Further, it compiles intersecting urban, environmental, economic, and social issues to enhance the infrastructure of sustainability transformation, and to overcome the challenges

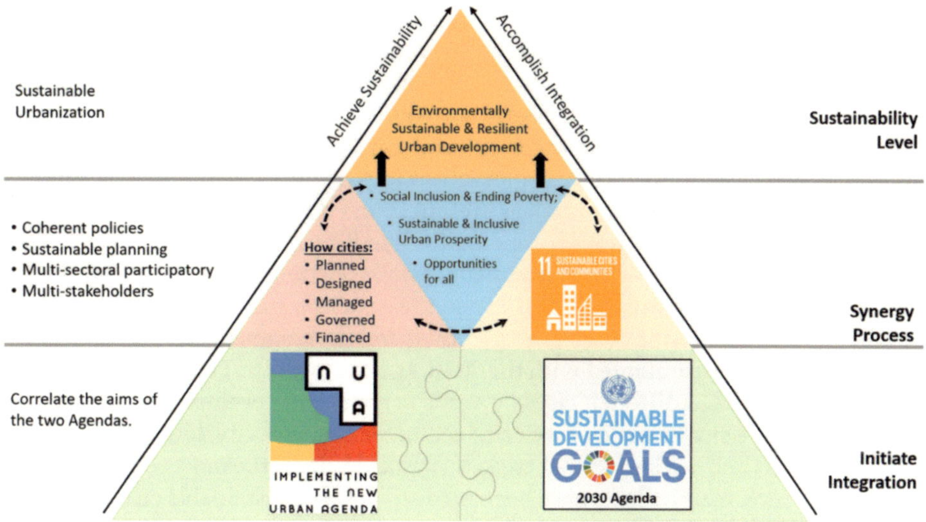

Figure 5.
The integration approach for the two agendas.

and obstacles that hamper human development by implementing well-built actions on sustainable planning and designing, ensuring mobilization of financial resources, governance, and sustainable management.

The first step (**Figure 5**) is to establish the integration basis based on the aims of the two Agendas, where they intersect in promoting urban sustainability. Then, enhance the synergy process through coherent policies, sustainable planning, and multi-sectoral and multi-stakeholder participation. While SDG 11 requires an integrated approach to urban development to overcome sectoral boundaries through building internal institutional synergies, building partnerships, and enhancing existing institutions [12]. This will make SDG 11 drive the NUA vision concerning urban planning, designing, management, governance, and finance. As a result, it will provide opportunities for all and, consequently, attain social inclusion, sustainability, and prosperity to end poverty. Eventually, the aim is to reach the sustainability level and accomplish sustainable integration between the two Agendas towards achieving sustainable urbanization in the urban agglomerations. Therefore, the integration of both Agendas will offer a robust base across different aspects. Further, a paradigm shift is necessary to ensure substantial and sustainable transformation of urban policies, planning, development, and management at all levels through SDG 11 within a comprehensive approach including entire SDGs.

5. Best urban sustainability practices

In 2007, the World Bank launched the Global City Indicators Program (GCIP) to deliver a platform for cities to analogize their indicators and exchange outcomes and best practices for sustainable urbanization [49]. A large area of cities' land is residential areas, while urban planning practices have not yet been properly incorporated [28]. The NUA realizes that there is a relationship between sound urban practices and development and urbanization [25]. Sustainability practices still lack interdisciplinary cogitation, enforcement, and inclusive decision-making procedures. While, some sustainability initiatives are under the environmental departments' responsibility [50]. Currently, economic and demographic considerations dictate the planning and design practices in urban development [51], and most sustainable urban development practices focus mainly on energy, waste, water, and transportation aspects, while the conservation of the environment and socio-cultural aspects are ignored [42, 52]. Whereas improper urban development planning practices impact the surrounding environment adversely [53], both policy and practices present sound direction to aspirational and normative sustainability and are appropriate for handling new and critical challenges [41], and link the economy and environment with the citizens' prosperity [54]. Urban areas are hotspots for attaining the SDGs, and the development of smart advanced technologies and solutions in the urban planning and development field is leading to a better sustainable future aligned with the 2030 Agenda (refer to **Table 1**) [45, 55].

The literature stated different city keywords in the urban development research, such as sustainable city, smart city, smart-sustainable city, eco-city, low-carbon city, resilient city, and knowledge city; whereas, sustainable city is the most common keyword [72]. Where, many efforts have been made to define a sustainable city based on an advanced normative proposition and an inventive empirical practice, and these efforts

Urban aspects	Best sustainability practices	References
Quality of life	Social interactions and networking; facilitate access to a wide range of services; maintenance and improvement of infrastructure	[56, 57]
Energy consumption	Use of renewable energy; adopt green building design technologies; adopt efficient management of energy resources; use energy-efficient technologies	[58–62]
Water consumption	Water conservation; sewage treatment; minimize water consumption; utilize the gray water; adopt smart techniques for wastewater treatment; employ flexible water distribution systems	[59, 63, 64]
Waste management	Minimize waste generation; waste separation and recycling; zero waste	[59, 65, 66]
Transportation	Encourage the use of public transport systems and expand; smart urban mobility; encourage walking and use of bicycles; adopt sustainable transport policies	[35, 62, 67]
Green economy	Adopt clean technologies; green tax policies; green infrastructure; circular economy	[63, 68]
Public space	Safety and security of public places; cultural and natural heritage protection	[59]
Biodiversity—flora and fauna	Increasing natural areas; improving landscape permeability; combat deforestation by law; making environmentally green spaces (e.g., gardens and parks); reinforcing urban nature conservation; protecting of ecological value of flora and fauna in their *situ*; managing urban expansion to avoid losses of sensitive biodiversity *situ*	[4, 21, 56, 69, 70]
Heat island	Increase vegetation cover; adopt actions to mitigate climate change; increase resilience; sustainable housing policies	[4, 15, 35, 56]
Land use and urban planning	Adopt green building design technologies; create environmentally sensitive areas to restore park and greenway systems; green infrastructures; integrate the resilience strategies with development and planning policies; protect the land instead of exploiting it	[35, 41, 56, 71]

Table 1.
Best urban sustainability practices.

led to systematically and technically defining sustainable urban development out of unified standardization, of which SDG11 is the newest and highest example [36, 73]. Therefore, working on sustainability goals is a complex task that needs high coordination and collaboration at multiple levels [74], thus, to accomplish long-term sustainability goals needs to develop and implement effective, measurable sustainability policies and practices that transform societies towards sustainable urban agglomerations [75].

6. Sustainable urbanization lessons in the context of SDGs

Urbanization is a multi-dimensional, multi-disciplinary, and cross-sectoral issue; it intertwines with politics, economics, society, governance, and the environment

SDGs	Lessons learnt
3 GOOD HEALTH AND WELL-BEING	• Access to safe drinking water • Develop and enforce integrated healthcare services • Avoid waste accumulation • Provide quality and affordable healthcare services • Eliminate health risks from air and water pollution • "Leverage health professionals to build public support for healthy city action across all sectors" [70] • "Build institutional linkages between health and other sectors, and promote norms of collaboration and cooperation through directed communication" [70]
6 CLEAN WATER AND SANITATION	• "Adopt integrated water management strategies, jointly addressing water supply, sanitation, and stormwater management" [70] • Assess water needs and future availability.
7 AFFORDABLE AND CLEAN ENERGY	• Control energy-consuming methods • Provide energy infrastructure for renewable energy production • Invest in the clean renewable energy sources, such as solar energy.
11 SUSTAINABLE CITIES AND COMMUNITIES	• Adopt sustainable-smart cities to combat the negative impacts of urbanization. • The increased number of vehicles led to more fuel consumption and more air pollution that contributes to climate change. Thus, need to invest in transportation-sustainable alternatives, such as electrical cars, public transportation, railways, and alternative fuel (e.g., biodiesel, methanol, and ethanol). • Provide affordable fees to public transport networks to encourage people to utilize them. • Provides education and research, health and social services, and cultural activities. • Efficient land utilization and provide basic services, such as water and electricity supplies. • To improve the economic conditions, make cities knowledge hubs to qualify and develop human resources skills, exchange innovation ideas, and pursue various careers. • Sustainable urban planning and management will provide a better social life, economy, and environment.

SDGs	Lessons learnt
12 RESPONSIBLE CONSUMPTION AND PRODUCTION	• Develop proactively integrated waste management strategies. • Encourage waste recycling investments.
13 CLIMATE ACTION	• "Cities can achieve net-zero emissions, but only if emissions are reduced within and outside of their administrative boundaries through supply chains, which will have beneficial cascading effects across other sectors" [77] • Protect water bodies, and increase public spaces and green spaces to contribute to mitigating climate change's impacts.

Table 2.
Some urbanization lessons learnt related to some SDGs.

sectors. Therefore, this issue is crucial and cannot be addressed independently due to its roots interlinked with multiple issues; such as climate change, pollution, resource consumption, infrastructure, economics, and other issues. For instance, the 2013 UN International Strategy for Disaster Risk (UNISDR) report stated that globally, cities are becoming increasingly vulnerable to severe environmental changes, such as drought, hurricanes, flooding, heat stress, heavy rainfall events, and other natural disasters (refer to **Table 2**) [76].

7. Conclusions

The two Agendas are more ambitious and nuanced and extend beyond the identification of matches between urban development needs and the aspiration mission of the SDGs, which require a concerted effort at all levels towards sustainability integration. The integration approaches are a way to move ahead; the most important is how to enact the integration approach between the two Agendas based on sustainability principles and practices. In many contexts, both Agendas offer a robust base for integration, which assists in including sustainability dimensions in urban planning development, including improving national urban policies and strategies, financial mobilization, adoption of urban governance, capacity-building, and institutional development. Overall, the holistic vision is the basis of the integrated approach to urban sustainability, which covers all sustainability aspects and a comprehensive understanding of the correlations among sustainability dimensions in both Agendas.

Despite the global attention to the significant role of cities, planning, and urban policy in shaping a typical sustainable city; there is a need to develop an approach to be a tool to simplify and resolve the complex and critical issues that hinder urban sustainability. Both Agendas are functioning to lead the way for the next decades' future because they have made it clangorous and clear that the world must drive on a sustainable path on multiple and a systemic track to ensure a better future. The policy tools are considered the cornerstone that needs to be calibrated in the context of the two Agendas to move forward in accomplishing sustainable urbanization.

The strengths of SDGs are interconnected and need integrated approaches; therefore, the urban planning and development process should embark on all aspects of the 17 SDGs or select the highest priority to ensure how agglomeration urban areas are structured, built, and planned and how services provided and run because these areas are fertile soil for SDGs implementation. Adopting a wide range of lessons and practices could assist the decision-makers and policy-makers in developing sounder actions and predicting potential issues for promoting future sustainable urbanization. Therefore, the key to success in attaining sustainable agglomerations is to redress the current course of action and put urban development on a sustainable path.

The key pass point for localizing the SDGs within urban development is the urban resilience strategies that reflect the obligations of global agreements and instruments. More specifically, the SDGs are a harmonious framework for NUA that allows the integration of diverse policy tools and strategies in a way that facilitates the sustainability localization process. To do so, it should focus on future urban planning, availability of resources (e.g., financial, technology, and data and information), stakeholders' involvement, knowledge and experience exchange, and collaboration, and capacity buildings for strengthening urban resilience. As a result, both Agendas will be localized at the national level in the realm of sustainable development.

To this end, we should more obviously realize that urban sustainability is an interdisciplinary endeavor to provide a unique framework to contribute towards accomplishing the SDGs in a more progressive global context in urban development and policy. Integration of the NUA and the 2030 Agenda could signify a moment for achieving sustainable urbanization that promotes sustainable and resilient urban futures.

Conflict of interest

The author declares no conflict of interest.

Author details

Abdulkarim Hasan Rashed
Independent Researcher in Environment and Sustainable Development, Bahrain

*Address all correspondence to: kme2001@hotmail.com

References

[1] Brundtland GH. Report of the World Commission on Environment and Development: Our Common Future. New York: United Nations; 1987

[2] UN. Rio Declaration on Environment and Development. Rio de Janeiro: United Nations; 1992

[3] Klopp JM, Petretta DL. The urban sustainable development goal: Indicators, complexity and the politics of measuring cities. Cities. 2017;**63**:92-97. DOI: 10.1016/j.cities.2016.12.019

[4] Sharifi A. Urban sustainability assessment: An overview and bibliometric analysis. Ecological Indicators. 2021;**121**:107102. DOI: 10.1016/j.ecolind.2020.107102

[5] UN-Habitat CIF. About the Cities Investment Facility. 2023. Available from: https://citiesinvestmentfacility.org/about-us/general/

[6] United Nations. Financing for Sustainable Development Report 2023: Financing Sustainable Transformations. New York: United Nations; 2023

[7] World Bank 2022. Available from: https://blogs.worldbank.org/ppps/mind-gap-time-rethink-infrastructure-finance; https://blogs.worldbank.org: https://blogs. worldbank.org/ppps/mind-gap-time-rethink-infrastructure-finance [Accessed: October 15, 2023]

[8] Almulhim AI, Cobbinah PB. Can rapid urbanization be sustainable? The case of Saudi Arabian cities. Habitat International. 2023;**139**:102884. DOI: 10.1016/j.habitatint.2023.102884

[9] Kosovac A, Acuto M, Jones TL. Acknowledging urbanization: A survey of the role of cities in UN frameworks. Global Policy. 2020;**11**(3):293-304. DOI: 10.1111/1758-5899.12783

[10] United Nations. World Urbanization Prospects. New York: United Nations Department of Economic and Social Affairs; 2018. Available from: https://population.un.org/wup/

[11] UN General Assembly (UNGA). Transforming Our World: The 2030 Agenda for Sustainable Development. 2015. Available from: http://www.un.org/ga/search/view_doc.asp?symbol=A/RES/70/1&Lang=E

[12] Vaidya H, Chatterji T. SDG 11 and the new urban agenda: Global sustainability frameworks for local action. In: Actioning the Global Goals for Local Impact: Towards Sustainability Science, Policy, Education and Practice. Singapore: Springer; 2019. pp. 173-185

[13] Xie L, Yan H, Zhang S, Wei C. Does urbanization increase residential energy use? Evidence from the Chinese residential energy consumption survey 2012. China Economic Review. 2020;**59**:101374. DOI: 10.1016/j.chieco.2019.101374

[14] Yu Y, Zhang N, Kim JD. Impact of urbanization on energy demand: An empirical study of the Yangtze River Economic Belt in China. Energy Policy. 2020;**139**:111354. DOI: 10.1016/j.enpol.2020.111354

[15] Anwar A, Sinha A, Sharif A, Siddique M, Irshad S, Anwar W, et al. The nexus between urbanization, renewable energy consumption, financial development, and CO_2 emissions: Evidence from selected Asian countries. Environment, Development

and Sustainability. 2022;**24**:1-21. DOI: 10.1007/s10668-021-01716-2

[16] Lu Y, Song W, Lyu Q. Assessing the effects of the new-type urbanization policy on rural settlement evolution using a multi-agent model. Habitat International. 2022;**127**:102622. DOI: 10.1016/j.habitatint.2022.102622

[17] Fang C, Yu D. Urban agglomeration: An evolving concept of an emerging phenomenon. Landscape and Urban Planning. 2017;**162**:126-136. DOI: 10.1016/j.landurbplan.2017.02.014

[18] Elimam H. Environmental problems & development sustainability in light of the Kingdom's 2030 vision: Opportunities & challenges. International Journal of Education and Social Science. 2022;**9**(1):2410-5171

[19] Acuto M, Ghojeh M. C40 cities inside out. Global Policy. 2019;**10**(4):709-711. DOI: 10.1111/1758-5899.12760

[20] Bellali J, Dellas E, Fischer K, Strauch L. Sustainable Development Goals, New Urban Agenda and the Urban Nexus. Berlin: Adelphi; 2016

[21] Ali-Toudert F, Ji L, Fährmann L, Czempik S. Comprehensive assessment method for sustainable urban development (CAMSUD)-a new multi-criteria system for planning, evaluation and decision-making. Progress in Planning. 2020;**140**:100430. DOI: 10.1016/j.progress.2019.03.001

[22] World Bank. Metadata Glossary [Internet]. 2018. Available from: Glossary | DataBank (worldbank.org) [Accessed: November 25, 2023]

[23] Fang C. Scientifically selecting and hierarchically nurturing China's urban agglomerations for the new normal. Bulletin of the Chinese Academy of Sciences. 2015;**30**(2):127-136. DOI: 10.16418/j.issn.1000-3045.2015.02.001

[24] Schindler S. The new urban agenda in an era of unprecedented global challenges. International Development Planning Review. 2017;**39**(4):349-355. DOI: 10.3828/idpr.2017.15

[25] UN-Habitat. New Urban Agenda. Nairobi, Kenya: United Nations Human Settlements Programme; 2017

[26] Sietchiping R, Reid J, Omwamba J. Implementing the SDGs and the new urban agenda. Environment and Urbanization ASIA. 2016;7(2):x-xii. DOI: 10.1177/0975425316660664

[27] Caprotti F, Cowley R, Datta A, Broto VC, Gao E, Georgeson L, et al. The new urban agenda: Key opportunities and challenges for policy and practice. Urban Research & Practice. 2017;**10**(3):367-378. DOI: 10.1080/17535069.2016.1275618

[28] UN-Habitat. International Guidelines on Urban and Territorial Planning. Nairobi: UN Habitat; 2015

[29] UN-Habitat. World Cities Report 2016: Urbanization and Development Emerging Futures. Nairobi: UN Habitat; 2016

[30] Beatley T, Brown J. Greening cities summary report. Global Platform for Sustainable Cities. 2022. Available from: https://www.thegpsc.org/sites/gpsc/files/deep_dive_learning_session_1_summary_report_11-3-22.pdf

[31] Zinkernagel R, Evans J, Neij L. Applying the SDGs to cities: Business as usual or a new dawn? Sustainability. 2018;**10**(9):3201. DOI: 10.3390/su10093201

[32] Valencia SC, Simon D, Croese S, Nordqvist J, Oloko M, Sharma T, et al. Adapting the sustainable development goals and the new urban agenda to the city level: Initial reflections from a comparative research project. International Journal of Urban Sustainable Development. 2019;**11**(1):4-23. DOI: 10.1080/19463138.2019.1573172

[33] Diaz-Sarachaga JM, Jato-Espino D, Castro-Fresno D. Evaluation of LEED for neighbourhood development and envision rating frameworks for their implementation in poorer countries. Sustainability. 2018;**10**(2):492. DOI: 10.3390/su10020492

[34] UN. Tracking Progress Towards Inclusive, Safe, Resilient and Sustainable Cities and Human Settlements: SDG 11 Synthesis Report High-Level Political Forum 2018. New York: United Nations Human Settlements Programme; 2018

[35] Croese S, Green C, Morgan G. Localizing the sustainable development goals through the lens of urban resilience: Lessons and learnings from 100 resilient cities and cape town. Sustainability. 2020;**12**(2):550. DOI: 10.3390/su12020550

[36] Caprotti F, Cowley R. Interrogating urban experiments. Urban Geography. 2017;**38**(9):1441-1450. DOI: 10.1080/02723638.2016.1265870

[37] Akuraju V, Pradhan P, Haase D, Kropp JP, Rybski D. Relating SDG11 indicators and urban scaling——An exploratory study. Sustainable Cities and Society. 2020;**52**:101853. DOI: 10.1016/j.scs.2019.101853

[38] Romero-Lankao P, McPhearson T, Davidson DJ. The food-energy-water nexus and urban complexity. Nature Climate Change. 2017;7(4):233-235. DOI: 10.1038/nclimate3260

[39] Steffen W, Richardson K, Rockström J, Cornell SE, Fetzer I, Bennett EM, et al. Planetary boundaries: Guiding human development on a changing planet. Science. 2015;**347**(6223):1259855. DOI: 10.1126/science.1259855

[40] Watson V. Locating planning in the new urban agenda of the urban sustainable development goal. Planning Theory. 2016;**15**(4):435-448. DOI: 10.1177/1473095216660786

[41] Elmqvist T, Andersson E, Frantzeskaki N, McPhearson T, Olsson P, Gaffney O, et al. Sustainability and resilience for transformation in the urban century. Nature Sustainability. 2019;**2**(4):267-273. DOI: 10.1038/s41893-019-0250-1

[42] Kaur H, Garg P. Urban sustainability assessment tools: A review. Journal of Cleaner Production. 2019;**210**:146-158. DOI: 10.1016/j.jclepro.2018.11.009

[43] Vela-Jiménez R, Sianes A, López-Montero R, Delgado-Baena A. The incorporation of the 2030 agenda in the design of local policies for social transformation in disadvantaged urban areas. Land. 2022;**11**(2):197. DOI: 10.3390/land11020197

[44] Nilsson M, Chisholm E, Griggs D, Howden-Chapman P, McCollum D, Messerli P, et al. Mapping interactions between the sustainable development goals: Lessons learned and ways forward. Sustainability Science. 2018;**13**(6):1489-1503. DOI: 10.1007/S11625-018-0604-Z/TABLES/2

[45] Blasi S, Ganzaroli A, De Noni I. Smartening sustainable development in cities: Strengthening the theoretical linkage between smart cities and SDGs. Sustainable Cities and Society. 2022;**80**:103793. DOI: 10.1016/j.scs.2022.103793

[46] Mtapuri O, Myeni SL. Sustainable development goals and the new urban agenda: A south African experience. In: Myeni SL, Okem AE, editors. The Political Economy of Government Subsidized Housing in South Africa. New York: Routledge; 2020. ISBN: 978-1-138-36491-2

[47] Rashed AH. The Impacts of Unsustainable Urbanization on the Environment. London, UK: IntechOpen; 2023. DOI: 10.5772/intechopen.110089

[48] Alberti A, Senese M. Developing capacities for inclusive and innovative urban governance. In: Cheema S, editor. Governance for Urban Services. Advances in 21st Century Human Settlements. Singapore: Springer; 2020. DOI: 10.1007/978-981-15-2973-3_6

[49] Zoeteman K, Mommaas H, Dagevos J. Are larger cities more sustainable? Lessons from integrated sustainability monitoring in 403 Dutch municipalities. Environmental Development. 2016;**17**:57-72. DOI: 10.1016/j.envdev.2015.08.003

[50] Garren SJ, Brinkmann R. Sustainability definitions, historical context, and frameworks. In: The Palgrave Handbook of Sustainability: Case Studies and Practical Solutions. Cham, Switzerland: Palgrave Macmillan; 2018. pp. 1-18. ISBN 9783319713892.

[51] Kumar A, Pushplata. Building regulations for environmental protection in Indian hill towns. International Journal of Sustainable Built Environment. 2013;**2**(2):224-231. DOI: 10.1016/j.ijsbe.2014.04.003

[52] Praharaj S, Han JH, Hawken S. Innovative civic engagement and digital urban infrastructure: Lessons from 100 smart cities mission in India. Procedia Engineering. 2017;**180**:1423-1432. DOI: 10.1016/j.proeng.2017.04.305

[53] Bai X, Nath I, Capon A, Hasan N, Jaron D. Health and wellbeing in the changing urban environment: Complex challenges, scientific responses, and the way forward. Current Opinion in Environmental Sustainability. 2012;**4**(4):465-472. DOI: 10.1016/j.cosust.2012.09.009

[54] Diaz-Sarachaga JM, Alvarez V. Assessment of the contributions of the Spanish urban agendas to achieving sustainable urban development. In: Conference: ICSD 2020 - International Conference on Sustainable Development; New York. New York: Royal Society of Chemistry; 2020

[55] Bibri SE, Krogstie J. Smart sustainable cities of the future: An extensive interdisciplinary literature review. Sustainable Cities and Society. 2017;**31**:183-212. DOI: 10.1016/j.scs.2017.02.016

[56] Dizdaroglu D. The role of indicator-based sustainability assessment in policy and the decision-making process: A review and outlook. Sustainability. 2017;**9**(6):1018. DOI: 10.3390/su9061018

[57] Ahvenniemi H, Huovila A, Pinto-Seppä I, Airaksinen M. What are the differences between sustainable and smart cities? Cities. 2017;**60**:234-245. DOI: 10.1016/j.cities.2016.09.009

[58] Stossel Z, Kissinger M, Meir A. Measuring the biophysical dimension of urban sustainability. Ecological Economics. 2015;**120**:153-163. DOI: 10.1016/j.ecolecon.2015.10.010

[59] Ameen RFM, Mourshed M. Environmental, social and economic challenges for urban development: Stakeholder's perception in a developing economy. In: Proceedings of the International Conference on Computing in Civil and Building Engineering;

Osaka. Japan: OSAKA University; 2016. pp. 6-8

[60] Verma P, Raghubanshi AS. Urban sustainability indicators: Challenges and opportunities. Ecological Indicators. 2018;**93**:282-291. DOI: 10.1016/j.ecolind.2018.05.007

[61] Chel A, Kaushik G. Renewable energy technologies for sustainable development of energy efficient building. Alexandria Engineering Journal. 2018;**57**(2):655-669. DOI: 10.1016/j.aej.2017.02.027

[62] Herrmann-Lunecke MG, Mora R, Sagaris L. Persistence of walking in Chile: Lessons for urban sustainability. Transport Reviews. 2020;**40**(2):135-159. DOI: 10.1080/01441647.2020.1712494

[63] Martínez R, Vela N, el Aatik A, Murray E, Roche P, Navarro JM. On the use of an iot integrated system for water quality monitoring and management in wastewater treatment plants. Water. 2020;**12**(4):1096. DOI: 10.3390/w12041096

[64] Tsegaye S, Gallagher KC, Missimer TM. Coping with future change: Optimal design of flexible water distribution systems. Sustainable Cities and Society. 2020;**61**:102306. DOI: 10.1016/j.scs.2020.102306

[65] Dizdaroglu D. Developing micro-level urban ecosystem indicators for sustainability assessment. Environmental Impact Assessment Review. 2015;**54**:119-124. DOI: 10.1016/j.eiar.2015.06.004

[66] Ng LS, Tan LW, Seow TW. Current practices of construction waste reduction through 3R practice among contractors in Malaysia: Case study in Penang. In: IOP Conference Series: Materials Science and Engineering. Vol. 271(1). Malaysia: IOP Publishing; 2017. p. 012039

[67] Garau C, Masala F, Pinna F. Cagliari and smart urban mobility: Analysis and comparison. Cities. 2016;**56**:35-46. DOI: 10.1016/j.cities.2016.02.012

[68] Zygiaris S. Smart city reference model: Assisting planners to conceptualize the building of smart city innovation ecosystems. Journal of the Knowledge Economy. 2013;**4**:217-231. DOI: 10.1007/s13132-012-0089-4

[69] Pierce J, Costadone L, Mannetti L, Morpurgo J, Green C, Halder M, et al. The Urban Nature Indexes (UNI): A new, flexible tool to comprehensively monitor urban ecological performance. Vol. Sep. 21. Research Square. 2023. pp. 1-20. DOI: 10.21203/rs.3.rs-3238847/v1

[70] Lee H, Siri J, Hasoloan J, Chapman TB, Das MB. Healthy Cities: Revisiting the Role of Cities in Promoting Health. Washington, DC: World Bank; 2023. Available from: http://hdl.handle.net/10986/40486

[71] Solly A, Berisha E, Cotella G. Towards sustainable urbanization. Learning from what's out there. Land. 2021;**10**(4):356. DOI: 10.3390/land10040356

[72] De Jong M, Joss S, Schraven D, Zhan C, Weijnen M. Sustainable–smart–resilient–low carbon–eco–knowledge cities; making sense of a multitude of concepts promoting sustainable urbanization. Journal of Cleaner Production. 2015;**109**:25-38. DOI: 10.1016/j.jclepro.2015.02.004

[73] Joss S. Sustainable Cities: Governing for Urban Innovation. London: Palgrave Macmillan; 2017

[74] Spiliotopoulou M, Roseland M. Urban sustainability: From theory influences to practical agendas. Sustainability. 2020;**12**(18):7245. DOI: 10.3390/su12187245

[75] Kanuri C, Revi A, Espey J, Kuhle H. Getting Started with the SDGs in Cities: A Guide for Stakeholders. New York: United Nations Sustainable Development Solutions Network; 2021

[76] UNISDR (United Nations International Strategy for Disaster Reduction). From Shared Risk to Shared Value–The Business Case for Disaster Risk Reduction. Geneva, Switzerland: United Nations Office for Disaster Risk Reduction (UNISDR); 2013

[77] IPCC (Intergovernmental Panel on Climate Change). Climate Change 2022: Mitigation of Climate Change. Working Group III Contribution to the Sixth Assessment Report of the Intergovernmental Panel on Climate Change. Cambridge, UK: Cambridge University Press; 2022

Section 2

Case Study Approaches

Chapter 4

Methodology for Indicator-Based Assessments of Environmental Hazards in Urbanised Areas: A Case Study of Ukrainian Regions

Kateryna Vasiutynska and Sergey Barbashev

Abstract

This study analyses the city and its surroundings as an integrated natural-anthropogenic system, emphasising ecological urbanisation within the global urbanisation framework. This component is represented as a cycle of interlinked adverse processes, delineating the resilience limits of natural systems under urban anthropogenic impacts. Utilising indicator-based methods, we assessed multifactorial urban impacts on the environment, proposing a practical approach to calculate the ecological urbanisation indicator. Using the indicator, we have investigated the regional aspect of the atmospheric environment's condition and identified the impact of urban sources of pollution. Additionally, we identified the main patterns of natural and human-induced hazards correlating with different levels of ecological urbanisation in different territories on the example of Ukrainian regions. Our approach addresses the uneven urban load across areas, which is essential for creating regional sustainable development strategies. It integrates environmental factors into regional safety management, requiring principles that regulate environmental impacts and promote safe land use.

Keywords: urbanised territories, indicator assessment methodology, environmental urbanisation, hazards, air resources in urban settings

1. Introduction

In today's world, it is not enough to define urbanisation as the growth and development of cities with a concentration of people who have chosen an urban lifestyle. Urbanisation represents a civilizational choice of humanity, a way to adapt to the conditions brought about by scientific and technological progress, which inevitably enhances the role of cities in societal development. The process of urbanisation is sweeping the globe: at the beginning of 2023, the urban population accounted for 57% of the world's total population, or 4.52 billion people, and according to data, this figure will gradually increase to 70% by 2050 [1].

Modern Ukraine ranks among the highly urbanised countries, with approximately 70% of its population residing in urban areas [2]. As of 2020, this placed Ukraine 76th in the global ranking of 195 countries and 25th among 44 European nations [3]. However, Ukraine exhibits several fundamental differences from other countries in terms of demographic processes. This is manifested in the sharp decline in the total population, including the urban population (depopulation rates were 1.35 for the total population and 1.2 for the urban population [1, 2] over the period 1991–2021). The coevolution of technological development (techno-genesis) and urbanisation processes has led to an extremely uneven distribution of the urban population (28.6937.08 million individuals as of January 2022 [2]) across 461 cities that occupy only 3% of the country's territory (19,000 km^2) [2]. The placement of Ukrainian cities corresponds to the economic complex established during the Soviet era, although their number has slightly increased (from 434 in 1989 to 461 in 2022 [2]). The population dynamics of cities have also shifted: 14 medium-sized cities (mostly mono-specialised) have been reclassified as small cities, and 2 cities (Donetsk and Dnipro) have lost their status as million-plus cities.

The environmental state of Ukraine's regions is formed by a combination of factors: on the one hand, the regional differentiation of technogenic (human-induced, technological) processes due to established economic complexes, and on the other, the high mosaic of natural landscapes and their varying types of resilience. The haphazard nature of urban development and the lack of strategic planning in alignment with sustainable development goals have led to a high degree of diversity between regions. This diversity is evident not only in levels of urbanisation (ranging from 37% in the Zakarpatsky region to 90% in the Donetsky region) but also in the ratio of small, medium, and large cities and other parameters indicative of "urban saturation" within administrative units. Furthermore, for most regions, the parameters of demographic urbanisation do not correlate with the indicators of city density and urban population density. Thus, the urbanisation process in different regions is uneven, with varying intensity and manifesting in diverse forms.

Urbanisation is driving negative changes in urban climate and landscapes, chronic pollution, and the dense location of hazardous industrial, energy, and transport facilities within or near residential areas. To detail the territorial features of cities' environmental impact as holistic entities, it is advisable to use the indicator method. Improving and further developing the system of environmental indicators requires taking into account the factors of homogeneity and differentiation of Ukraine's regions by the potential of urban hazards to the environment, human life, and health. Integrating ecological indicators into the urbanisation indicators system enables the identification of the impacts of cities, varying by population size and functional purpose, on environmental elements and also allows to differentiate administrative regions of Ukraine by the nature of such impacts. The practical application of the urbanisation indicator lies in determining the regional level of urban-generated pressure on territories and predicting their implications within the framework of the country's sustainable development plans.

2. Advancing the indicator method for assessing the ecological resilience of urban areas

An indicator is the most generalised measure of the state of the system and its dynamic changes, among which there are stand-alone and aggregate indicators.

Stand-alone indicators reflect specific aspects of the state of the system under study, while aggregated ones correspond to complex indicators. They reflect the weight of the indicator, the degree of its involvement in the assessment, the coefficient of use, and correspond to the term "index".

The concept of global ecological indicators, put forth by the Organization for Economic Co-operation and Development (OECD) in 2004 [4], concurrently provides a unique collection of environmental statistics for 10 indicators (Climate Change, Ozone Layer, Air Quality, Waste Generation, Freshwater Quality, Freshwater Resources, Forest Resources, Fish Resources, Energy Resources, and Biodiversity) [5]. This collection facilitates the evaluation of environmental policy effectiveness predominantly in European and some non-European countries. Regrettably, Ukraine's presence in the Environment Statistics database [5] is limited to certain sections (greenhouse gas emissions). The EU's strategy for the environmental dimension of Sustainable Development (SD) is based on indicator groups aligned with Sustainable Development Goals (SDG), among which "Sustainable Cities and Communities" indicators [6] outline European policy trends in this direction. Additionally, the UNECE Working Group on Environmental Monitoring and Assessment recommended 118 top-priority indicators for Eastern European countries from a basic set of 356 indicators based on seven criteria [7]. Current EU trends, as outlined in Eurostat's publication "Sustainable Development in the European Union — 2022 Monitoring Report on progress towards the SDGs in an EU context" [8], though not an official stance of the EU, reflect the main trends of changes in around 100 indicators structured around 17 SDG. Over a 5-year (short-term) period, among the "Sustainable Cities and Communities" indicators, 11 are aimed at increasing community access to basic services, energy, housing, transportation, and green spaces, while simultaneously reducing resource use and environmental impact.

Using various indicator systems and indices to assess environmental conditions in the light of accelerated urbanisation processes sweeping many countries across Asia, America, and Europe is a "top topic" of scientific research. The rapid urban development and significant population growth in China have spurred numerous studies assessing air pollution in urbanised territories [9–11], climate changes [12], shortages of quality drinking water [13, 14], and other issues linked to the growing imbalance between the natural environment and urbanised areas.

The sustainable use of water resources is a crucial prerequisite for the sustainable development of urban areas and countries in general. Urbanisation processes have a comprehensive impact on water bodies not only near cities, but also far beyond them. Global urbanisation and environmental degradation are leading to a rapid decline in the quality and quantity of water resources, thereby limiting their suitability for comprehensive use. Huge pressure on water resources from an expanding population is associated with changes in land use [14, 15], development in watershed areas, and alterations in vegetation cover [13], as well as groundwater depletion in the world's megacities [16].

Numerous indicators and indices reflect the importance of combining integrated urban water management and planning for increasing water consumption over time. The Driving Forces, Pressures, States, Impacts, and Responses (DPSIR) framework facilitates the assessment of water use indicators according to sustainable development criteria [17]. The DPSIR indicator system can be presented as a set of indicators reflecting the ecological status responses of urban systems to anthropogenic pressures of various genesis. DPSIR indices are determined based on statistical data and are part of the well-known Pressure-State-Response (PSR) indicator system [18, 19], developed by the OECD [20].

The shortage of fresh surface water transforms the water management issue into a national threat for many countries, including Ukraine. Sustainable water use indicators [21] compare demographic-economic factors (population growth, gross domestic product) with water demand and water availability. The use of such a system of indicators justifies the regulation of demography and the economy, ensuring compliance with clear rules to limit pollution of basin areas, preserve aquifers, and save water.

The value of the indicator approach in identifying correlations between economic and ecological subsystems lies in its integration into practices across various management levels [22–24]. Among the key indicators assessing the effectiveness of environmental policy, ecological taxes are highlighted as crucial in Ref. [25], based on indicators that are regularly monitored by the European Environment Agency (EEA), OECD, and EUROSTAT. The contemporary perspective on environmental regulation encompasses the integration of urban land use efficiency, green technology innovations, and the conservation of both natural landscapes, especially forests, and urban green spaces [15, 26–28].

Ukraine ranked 36th out of 165 countries in the Sustainable Development Index, which is annually compiled by the United Nations and the Bertelsmann Foundation ([29], p. 470). Unfortunately, the performance indicators for SDG11 – Sustainable Cities and Communities are currently in the "orange zone" and do not yet reflect the country's progress towards sustainable urbanisation. Ukraine has joined the global effort to achieve sustainable development, with strategic frameworks set for the period up to 2030 [30]. The national system consists of 86 national development goals and includes indicators to monitor the realisation of goals and targets, adapted to the Ukrainian context. Ukraine has presented three voluntary National Monitoring Reports: "Sustainable Development Goals: Ukraine 2019" [31], "Sustainable Development Goals: Ukraine 2020" [32], and "Sustainable Development Goals: Ukraine 2021" [33]. The war in Ukraine has interrupted not only the nascent monitoring process but also the country's development itself. However, Ukrainian scientists continue to work both on improving indicators and on methods of their measurement [34] and on analysing the reasons that did not allow achieving the Goal 11 (Sustainable development of cities and communities) [35].

Overall, the implementation of a sustainable development strategy and a voluntary reporting system in Ukraine has stimulated the development of comprehensive indicators for assessing the environmental condition of regions and the sustainability of individual cities, including indicators of the environmental dimension [36–42].

Unfortunately, Ukraine still lacks sufficiently developed indicator systems for assessing the regional aspects of sustainable development. Research by national scientists using the index approach has primarily focused on comparing the country's administrative regions based on natural and technogenic hazard factors [43–47]. The decrease in the total number of natural disasters and the 1.3-fold excess of man-made threats over natural ones in most regions of the country are established in Ref. [45]. The advantages of the indicator method include the broad application of mathematical models and modern information technologies, which enable not only the assessment but also the prediction of territorial risk conditions [47]. In general, the indicators are intended to be the basis for the implementation of coordinated measures to minimise and prevent negative environmental changes and deterioration of overall safety. Their implementation into Ukraine's actual policy should aim to address global issues such as poverty alleviation, ensuring high quality of life, and the safety of urban populations.

In contrast, our analysis of methods for ranking Ukrainian regions based on ecological safety [48] revealed significant discrepancies in the assessment results obtained using indexes and risk indicators. This contradicts the principle of systematisation and diminishes the effectiveness of ecological safety management at the national level. Regional "safety" rankings do not highlight the contribution of urban systems to the ecological state of a region. Generally accepted assessments are limited to the criteria of unilateral negative impacts of cities on the environment and do not include reverse cumulative effects when areas are globally covered by urbanisation processes.

3. The formation of the environmental state of the "city-adjacent territories" system

To accurately consider the "city – adjacent areas" system, let us clarify some concepts. Indeed, it is crucial to differentiate between "city" and "urban agglomerations," as well as "suburban areas" and "adjacent areas". In Ukraine, the term "suburban territory" is not clearly defined ([49], p. 84). It can refer to city green belts, like forests and parks, as well as small suburban settlements. It also includes urban high-rise districts, such as the "Rainbow" area in Odesa. Additionally, farms, agricultural lands, and industrial enterprises that create urban industrial zones fall under this category. Various interpretations of "suburban zone" are discussed in Ref. [50]. Indeed, the interconnection between a city and its suburban areas is in a symbiotic development. Unfortunately, in Ukraine, the features of suburban areas' development are almost unstudied, except for the general acknowledgment that they are impacted by cities and slowly become urbanised. Official information sources do not include data on city territory expansion through the incorporation of suburban areas. However, the problem of merging cities with suburban areas into urban agglomerations is of utmost importance all over the world. The main stages of this process for different continents are analysed in Ref. [51]. The authors noted terminological discrepancies in defining urban agglomerations, identified infrastructure expansion (especially railway lines) as a leading factor in urban space growth and aptly described this process as "Urban sprawl flooding".

Urban agglomerations serve both as a reserve for city development and as a framework for urban settlement within administrative-territorial units. Unfortunately, a legislative definition of urban agglomeration in Ukraine (23 ones by official counts as of the beginning of 2013) is still lacking. The cities within the 19 largest agglomerations were home to about 17 million people, accounting for 36% of the country's population [2]. However, the State Strategy for Regional Development for 2021–2027 [52] only recognises following agglomerations—Kyiv, Kharkiv, Dnipro-Kamianske, Zaporizhzhia, Kryvyi Rih, Odesa, and L'viv. It is important to note that the World Bank (OECD) [53] establishes these agglomerations using a publicly undisclosed methodology, which differs from the night-time lights (NLS) mapping method used.

Since 2017, the draft law "On Urban Agglomerations" [54] has yet to be considered in Parliament. In 2022, the concept of a draft law defining "Agglomeration as a special form of cooperation between territorial communities (urban, settlement, rural) located in the zone of impact of the agglomeration centre" was developed by the Association of Ukrainian Cities [55]. The revised edition emphasises that agglomerations are essential for both the communities and the state, and that formalising this cooperation model in Ukrainian legislation will create additional opportunities for the development of territories and Ukraine's economy. In September 2022, a public discussion on the role of agglomerations in war and post-war reconstruction took place in L'viv during the IX National Forum of Local Self-Government, supported and moderated by the Council of Europe [56]. Regrettably, since the onset of the war, this law has not been adopted. The delay in the legislation creates a significant gap in utilising the potential of large cities in Ukrainian regions. In Ref. [57], the adverse effects of the lack of legal regulation on local self-government development are outlined. In Ref. [58], the focus is on logistical connections and the creation of economic and infrastructure clusters, as well as the degree of cooperation in production processes and services. The adoption of the law on urban agglomerations would enable harnessing the potential and capabilities of large cities to enhance the competitiveness and economic growth of the respective regions and the country as a whole.

Thus, we delved into the nature of relationships within the "city – adjacent territories" system model, examining it as a nexus of interconnected interactions [59]. A city is considered within its administrative-territorial boundaries as defined in Ref. [2] and represented as mandatory statistical information in the Environmental Passports of regions [60]. Adjacent territories are those that surround a city, potentially including or extending beyond suburban areas. The rapid degradation of territorial resources and alterations in natural cycles extend far beyond urban settlements. Areas impacted by urbanisation are continually evolving under chronic anthropogenic pressure, significantly altering natural processes. Key factors of urban-technogenic impact include the following:

Impact on Water Resources: anthropogenic changes in ecosystems in catchment areas (e.g. deforestation, drainage of marshes, eutrophication of water bodies); development, fragmentation, pollution of catchment areas; changes in land use patterns; changes in the hydrological regime due to excessive consumption, climatic factors, construction of hydraulic structures; changes in the hydro-chemical regime as a result of wastewater discharge [61].

Impact on Airspace (impact of "city breathing"): pollution from mobile and stationary sources, including emissions of greenhouse gases and ozone-depleting substances; transformation of quantitative and qualitative parameters of the atmosphere, initiation of natural disasters (hurricanes, whirlwinds, tornadoes, etc.) related to climate change [62].

Impact on land resources: changes in the hydro-geological regime; mechanical loading on the soil due to infrastructure objects (roads, pipelines) [43]); waste disposal; development of hazardous geodynamic processes (karsts, landslides, flooding, etc.), in addition to the changes in the structure of land use.

Impact on biota: biodiversity loss, loss or critical decline in population numbers, destruction of habitat, disruption of wildlife migration corridors, impacts on fish spawning by hydraulic structures.

Impact on environmental safety: consequences of man-made accidents, man-made fires, and initiation of forest fires [63]; initiation of dangerous exogenous geological processes (landslides, mudflows, karst, flooding); initiation of natural disasters due to urban climate changes.

All groups of factors have a cumulative impact on natural landscapes, which are initially formed under these effects. It should be noted that changes in the land use structure have a direct impact on water and land resources, and an indirect impact on atmospheric (through changes in the conditions of dry and wet deposition of pollution) and biological (through changes in habitat conditions) ones.

The relationship between urbanisation and changes in the ecological state of most Ukrainian cities and their adjacent areas cannot be simplified into broad generalisations. The deterioration of the quality of the components of the natural and built environment complicates multi-level natural and anthropogenic phenomena. Adverse events and processes, when interacting within landscapes of different types, create environmental situations with varying degrees of tension. The ability of natural landscape components (natural water bodies, forest and steppe areas, comprehensive green zones) to maintain their core resilience parameters under anthropogenic pressure is key to minimising natural-technological adverse events in urban areas.

Urban sprawl and the formation of urban agglomerations lead to the expansion upon suburban natural areas, change the way natural resources are utilised, and lead to the loss of their ecosystem services [64, 65]. Overall, this disrupts the stability of natural complexes, triggers, and intensifies hazardous phenomena, which, in turn, adversely affect the state of ecological risk in urban systems.

There exists a direct and back relationship between the ecological state of the urban territories and the extent of transformation and degradation of natural systems in adjacent territories, where anthropogenic pressure on suburban natural areas initiates a certain sequence of adverse events, schematically represented in **Figure 1**.

The diagram illustrates a cycle of unidirectional adverse phenomena and processes that disrupt the balance of the "city – adjacent areas" system through a mechanism of positive feedback:

- the city determines degradation and destructive processes in the adjacent territories, increases the overall potential of hazards, and intensifies their negative consequences;

- as a result, significantly altered and distorted natural processes accelerate degradation and increase the level of danger in the urban environment.

Thus, the boomerang effect from the initiation of this cycle (**Figure 1**) returns to the urban system. The city not only generates urban-technological impacts on the entire urbanised area but also evolves within the ecological situation and circumstances created by such impacts. Changes in the quantity and quality of resources (biological, mineral, climatic, landscape) lead to increased demand for them. This dictates the extensive development of cities, counteracting the goals of sustainable

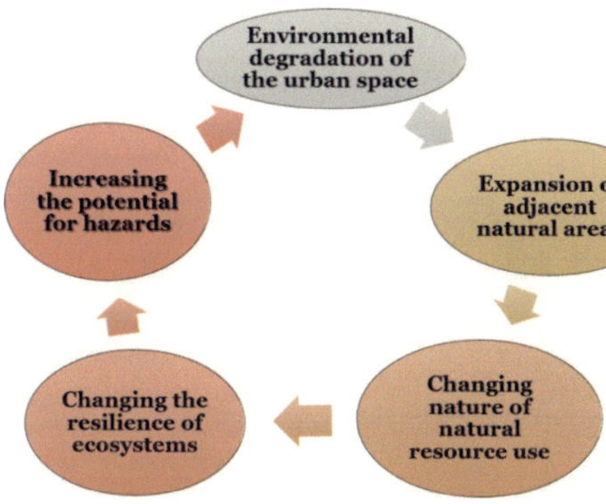

Figure 1.
The positive feedback loop in the "city-adjacent territories" system.

development. Therefore, urban communities should be interested in limiting urban-technological impacts on the environment and developing cities in an intensive scenario in line with the SDGs.

The analysis of the city-adjacent territories system linkages essentially provides a theoretical basis for urban ecological resilience [66], encouraging the revitalisation and change of urban areas and cities to improve living conditions, promote innovation, and reduce environmental impact, maximising economic and social co-benefits.

4. Understanding ecological urbanisation

Traditionally, urbanisation is viewed in terms of transitioning to an urban lifestyle. Regional urbanisation processes are considered through the settlement system in large, medium, small cities, and megacities. Today, urbanisation is an all-encompassing process that affects all facets of life. Demographic changes correspond to the nature of a country's post-industrial development but do not reflect all the interconnected aspects of the urbanisation process, components of which are presented in **Figure 2**. It should be noted that the concepts of demographic, spatial, and socio-economic urbanisation have already been utilised in Ref. [24]. We proposed from Refs. [59, 67] a set of interdependent components of the urbanisation process (**Figure 2**), which collectively determine the ecological state of territories. Among them, only demographic urbanisation is quantitatively assessed through the percentage of urban population. The nature of the impact of each component has both

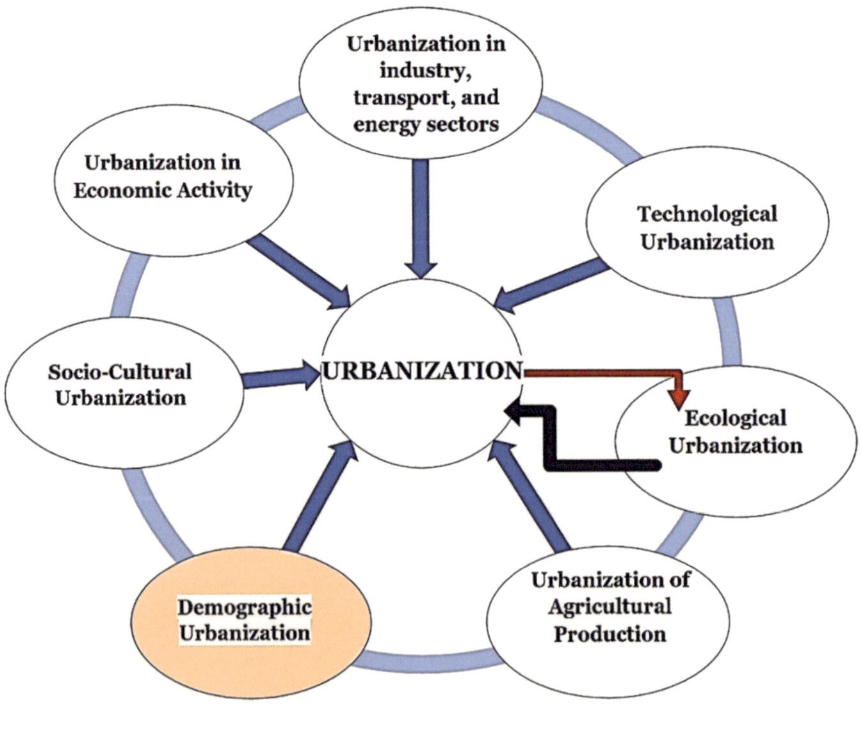

Figure 2.
Framework of urbanisation process elements.

negative and positive sides. For example, while industrial urbanisation is directly associated with negative environmental pollution, climate and landscape changes, and other destructive processes, technological urbanisation, through structural and technological modernisation of the industrial sector of the economy, mitigates the effects of industrialisation. With the modern production of complex goods (machinery, equipment, tools) and information services (patents, licenses, software products), the pressure on all environmental components is reduced compared to the waste generated by the extraction and processing industries.

Urban enterprises reap positive effects from urbanisation due to the concentration of labour, financial, material, and natural resources, intensive information exchange, and improved conditions for innovation and investment. Infrastructure costs are reduced due to the localisation of industrial zones and the formation of agglomerations, among other factors. Socio-cultural urbanisation accompanies and partly conditions the scientific and technological evolution predominantly of cities. The need to meet the growing demands (food, water, raw materials, etc.) of urban residents stimulates the urbanisation of agricultural production. The growth of cities requires sustainable land management, a move away from extensive forms of agricultural development, and the adoption of modern environmentally friendly production methods for all types of agricultural products. The urbanisation of economic activity lays the groundwork for stimulating innovative development in the regions of Ukraine [68], facilitating the use of modern eco-economic tools in implementing sustainable development programmes and managing the safety of the urban environment.

Under ecological urbanisation, we comprehend the totality of negative processes and phenomena, specifically the degradation and depletion of natural resources, landscape losses, changes in ecosystem productivity and biodiversity, and alterations in natural cycles. These occur not only within cities but also extend far beyond urban settlements and define the degree of transformation of natural systems under the pressure of urban-technological impacts. Ecological urbanisation may coincide with spatial urbanisation and be measured as the proportion of urbanised area relative to the total area of a region or country. Essentially, this quantitative indicator reflects the extent of artificial surface within the urban area.

Artificial surfaces can be represented by impervious surfaces due to changes in soil mechanical properties (compaction, loosening, weathering) and disruptions in gas and water exchange, or by completely sealed (isolated) surfaces made of asphalt and concrete coverings of streets and roads, roofs, buildings, industrial areas, and disturbed lands. In both cases, the artificial surface of urban space does not participate in the natural cycle. Therefore, destructive processes and adverse natural phenomena (e.g. disruptions in water runoff regime, and hydrogeological and geological hazards [68]) lack mechanisms for natural regulation and self-recovery. It is the artificial surface that is responsible for the creation of heat islands in cities and other factors of climate change [69]. As a result, the global stability of the territory is disrupted. The spread of artificial surfaces within cities and urban agglomerations is associated with transport infrastructure, construction, and changes in land use.

Within the system of urbanisation process components, it is important to note the dualism of ecological urbanisation (indicated by black and red arrows in **Figure 2**). On one hand, environmental elements and biological populations (including humans) are the primary recipients of the multifaceted impact of all urbanisation components (the black arrow in **Figure 2** indicates this impact). It is this impact that causes degradation and disrupts the stability boundaries of the "city-environment" system. On the other hand, ecological urbanisation acts as a limiting factor in all evolutionary changes due to the limited availability of exhaustible natural resources, energy sources, and land resources. Anthropogenic changes in the quality and quantity of natural resources and the scope of their ecosystem services create natural barriers to socio-economic, scientific, and technological development of society, and slow down the tempo of urbanisation.

To preserve the "ecological health" of urban systems, establishing ecological boundaries is crucial, dictated by the imperative to conserve part of the ecosystems, the so-called "wild nature," beyond anthropogenic impact, to maintain the global stability of territories. Thus, ecological urbanisation, through the limitation of natural and territorial resources, restricts the intensity of the entire urbanisation process (indicated by a reverse red arrow in **Figure 2**). Therefore, ecological urbanisation is a defining factor in the stability of the "city – adjacent areas" system and a limiting factor in the urbanisation development of territories.

5. Developing an indicator approach to identify urbanisation characteristics in Ukrainian regions

We suggest using an indicator of ecological urbanisation to evaluate the environmental health of administrative units. As previously discussed, ecological urbanisation is defined by the scale of urban space (the city and its adjacent areas under its direct and indirect influences) and depends on factors such as the proportion of urbanised territory and the density of the urban population, which affect the intensity of natural resource use and determine the nature of land use. This approach does not require a large number of indicators or specialised methods for their processing. Simultaneously, the selection of indicator parameters aligns with fundamental ecosystem laws. All data presented below are based on indicators as of the beginning of 2022 [2, 67].

The ecological urbanisation indicator ($I_{ec.urb}$) was calculated as a linear combination of normalised indicators of urbanised population density ($\rho_{pop.urb}$) and the proportion of land occupied by urban settlements (S_{urb}) relative to the total area of the region, as demonstrated in Ref. [67]. The indicator (S_{urb}) is assumed under the premise that the entire city territory is conventionally artificial. In addition to the sealed surface of residential and industrial areas, even urban green spaces are mostly artificially created systems in which natural processes are modified and subordinated to anthropogenic activities (organisation of plant supervision, selection of species composition, soil cultivation, removal of leaf litter, etc.). In our view, the indicator of the proportion of land occupied by cities relative to the total area of the region reflects the overall level of urban coverage of the territory and allows, subsequently, to define the ecological boundaries of urbanisation.

The ranking of Ukrainian regions by $I_{ec.urb}$ indicator ranges presented in Ref. [67] allowed us to organise administrative territories of different functional purposes and varying degrees of population aggregation by the level of urban saturation. It was found that the Donetsk region significantly differs from others, characterised by the highest value of the $I_{ec.urb}$ indicator. Other regions are conditionally divided into four groups based on the index ranges. We also conducted a cluster analysis of the $(I_{ec.urb})^n$ values normalised by the standard procedure [63] using the k-means method [70] to determine the affiliation of Ukrainian regions to a single group, in which the regional characteristics of urbanisation processes are more similar compared to entities from other groups. We identified three clusters:

I. Indicator values $(I_{ec.urb})^n$ close to the median (0.37) (numbers in the interval between the first and third quartiles); 15 regions (excluding AR Crimea, for which there is no data): Cherkasy (0.315), Sumy (0.320), Zaporizhzhia (0.355), Kherson (0.358), Mykolaiv (0.365), Rivne (0.375), Chernivtsi (0.390), Odesa (0.400), Ternopil (0.405), Vinnitsa (0.425), Khmelnitsky (0.450), Ivano-Frankivsk (0.470), Volyn (0.500), Luhansk (0.500), and Kharkiv (0.505);

II. Indicator values $(I_{ec.urb})^n$ below the median (below the first quartile); 6 regions: L'viv (0.165), Zakarpattia (0.230), Chernihiv (0.240), Zhytomyr (0.265), Kirovograd (0.270), and Poltava (0.300);

III. Indicator values $(I_{ec.urb})^n$ above the median (above the third quartile); 3 regions: Dnipropetrovsk (0.540), Kyiv (0.590), and Donetsk (0.775).

This differentiation does not align with the traditional division into southern, northern, eastern, and western macro-regions but reflects the general characteristics of urbanisation processes. It should be noted that Ukraine, as the content of cluster I shows, is a medium-urbanised country with a disproportion of types of demographic settlement in medium and small cities. Low-density urban population living in urban settlements, which occupy a small portion of the total regional area, characterises cluster II. Cluster III, consisting of only three regions, is distinguished by a combination of maximum urban population density and significant volumes of urbanised territories in the regions.

The regional differentiation in the intensity of urbanisation processes has led to significant differences among regions in terms of demographic behaviour, which correlates with the index of actual urbanisation [71]. To account for the settlement systems of

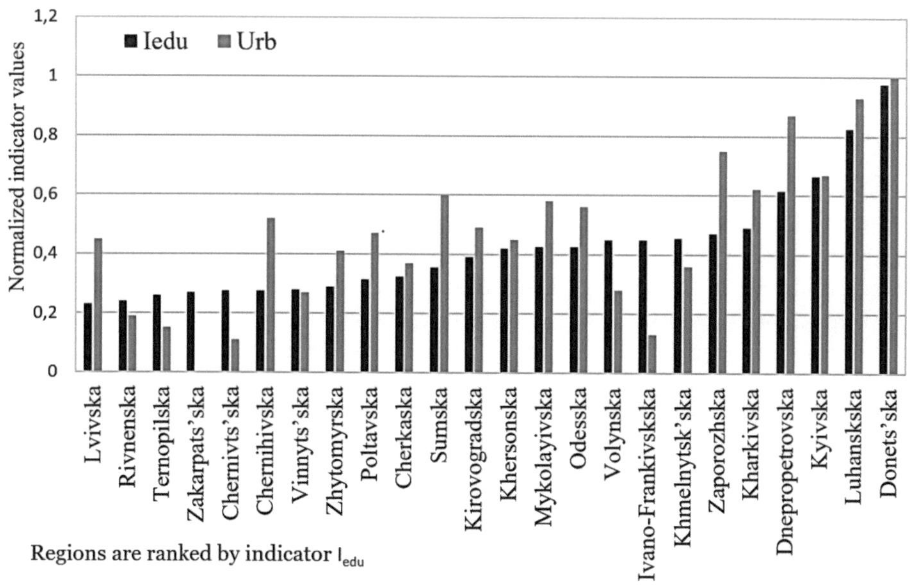

Figure 3.
The value of the ecological-demographic urbanisation index in relation to the level of demographic urbanisation in Ukrainian regions.

different regions, the indicator of real urbanisation ($I_{real.urb}$) was applied. It was calculated as the geometric mean of four indicators: X1, the proportion of urban population in the total population of the region (the region's level of urbanisation); X2, the proportion of urban settlements in the total number of settlements; X3, the proportion of cities with a population of more than 50,000 in the urban population; X4, the proportion of cities in the total number of urban settlements [72]. A complex index of ecological-demographic urbanisation (I_{edu}) was calculated based on a linear combination of real and ecological urbanisation indicators. Its values, in comparison with the normalised indicator of demographic urbanisation level (D) for the country's regions, are presented in **Figure 3**. Thus, the integral indicator, the ecological-demographic urbanisation index, takes into account both demographic and ecological factors of urbanisation processes.

Indeed, the proposed indicators are based on a limited set of readily available metrics, straightforward to calculate, which can be employed to distinguish Ukrainian regions by their urban load. We will show how we used these indicators to study the state of natural environment elements and patterns of natural-technogenic threats depending on the level of ecological urbanisation in the regions.

6. Characteristics of atmospheric pollution across regions of Ukraine with diverse urbogenic loads

An ecologically safe atmospheric environment is an indispensable condition for ensuring high quality of life and socio-economic activity of people. Transformations in the quantitative and qualitative parameters of the atmosphere are inextricably linked to climate changes, as they share a common source: energy generation and carbon fuel combustion [9, 10, 22, 24, 62, 73–75]. Air pollution in densely populated urban agglomerations is not limited to local effects. Pollutants are transported over vast distances, up to thousands of kilometres, significantly contributing to the overall background atmospheric pollution. In general, the maximum negative effects of changes in the quality of the atmospheric environment are typical for the most urbanised industrial centres. According to the Pollution Index for European cities, annually reported by the online database numbeo.com [76], major regional centres in Ukraine such as Dnipro, Odesa, Kyiv, Kharkiv, and L'viv are among the 50 most polluted European cities, with Dnipro, scoring 80.98 on the Pollution Index, ranking sixth. Exceeding the World Health Organization (WHO) target values for the annual amount of fine particulate matter ($PM_{2.5}$), which causes approximately 50,000 premature deaths per year, is also common in urban areas of Ukraine. As of 2020 [77], Ukraine ranks 8th in Europe (with an annual average $PM_{2.5}$ concentration of 19.2 $\mu g/m^3$), and Kyiv ranks 39th among the capitals of 106 countries worldwide.

The analysis of the dynamics of pollutant quantity changes from stationary and mobile sources over the period 1995–2020 [78] has shown a positive impact of the urbanisation process on the reduction of emissions from stationary sources: the pollution load per capita in the country decreased by 1.8 times. In recent years, there has been an increase in the negative trend of environmental pollution by transport, resulting in a 1.13-fold increase for pollution per capita. This situation in Ukraine aligns fully with global trends.

The analysis of the ratio of stationary and mobile source capacities in regions with varying levels of urban impact (**Figure 4**) revealed the following trends: firstly, an increase in total emissions from both types of sources in line with the growth of the (I_{edu}) index; secondly, the prevalence of stationary emissions over transport

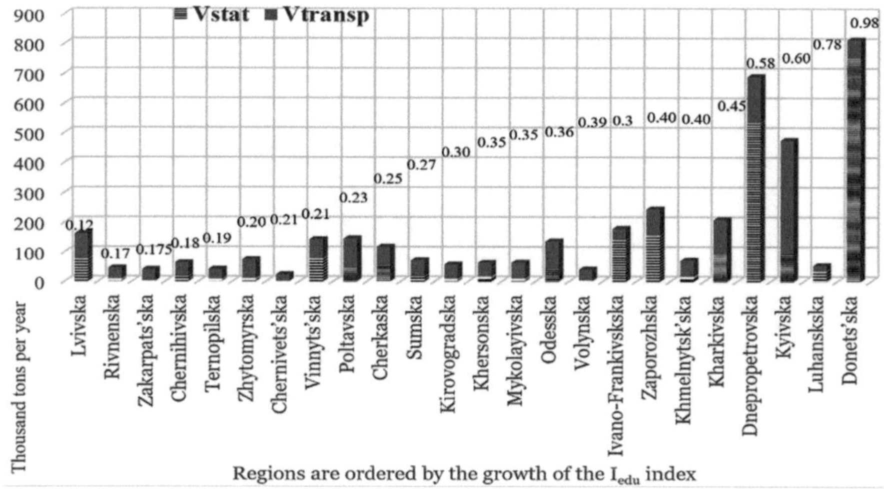

Figure 4.
*Emissions from stationary and transport sources by region of Ukraine compared to the I_{edu} index. *Kyiv region includes data for the city of Kyiv.*

emissions in only five regions—Vinnitsa, Dnipropetrovs'k, Donetsk, Zaporizhzhia, and Ivano-Frankivsk. In Dnipropetrovsk and Donetsk regions, they account for about 60% of the total emissions from stationary sources. Transport emissions dominate in most regions across various levels of ecological-demographic urbanisation, with the most significant impact observed in highly urbanised areas such as Dnipropetrovsk, Kharkiv, and Kyiv regions. The peak in the Kyiv region reflects the capital's impact on the sharp increase in vehicle traffic.

Evaluating the impacts of urbanisation faces the challenge of defining their boundaries. Polluting aerosols emitted through the high smokestacks of energy and industrial plants tend to spread over vast distances, significantly affecting the global safety of the atmospheric air. We have identified the contribution of urban stationary sources to the regional level of air threat through the coefficient of urban air pollution of regions $(Urb_{stat})^n$ (the part of emissions from urban stationary sources in the total regional capacity) and have shown a sufficient level of correlation with the indicator of environmental urbanisation $(I_{ec.urb})^n$ (**Figure 5**).

The hazard level of the air basin of cities of regional significance was also assessed accordingly $(I_{edu})^n$ [62]. Regardless of the urbanisation level, cities with a concentration of hazardous techno-sphere objects experience the highest environmental load, as demonstrated by the examples of Ladyzhyn, Cherkasy, and others. On the other hand, cities, as economic development centres, make better use of science and available technologies. In the management of large cities, cutting-edge technologies are more widely implemented, and modern air protection measures are better organised. Enterprises are typically located outside city borders, in industrial zones. Thus, the positive impacts of urbanisation can explain the low specific atmospheric pollution load per unit of area and population in cities of the Odessa, Kherson, Kyiv, and Khmelnitsky regions, which have a sufficiently high level of ecological-demographic urbanisation [62].

Our research emphasises the priority of combating transport pollution, which prevails in most regions regardless of the level of ecological-demographic urbanisation. Therefore, controlling emissions from mobile sources, urban road planning

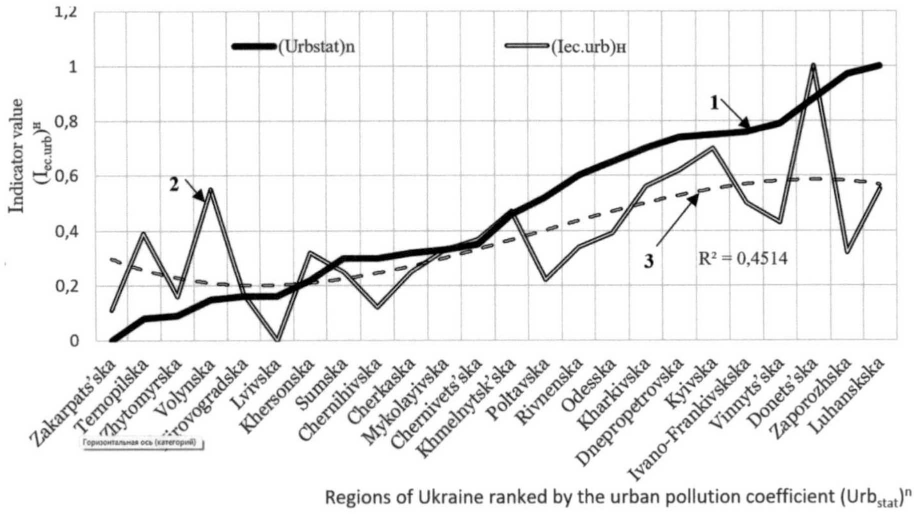

Figure 5.
Combination of indicators of air pollution by urban stationary sources and the indicator of environmental urbanisation: 1—$(Urb_{stat})^n$; 2—$(I_{ec.urb})^n$; 3—polynomial trend line for the $(I_{ec.urb})^n$ indicator.

and organising traffic flows in cities and beyond are necessary measures for a sustainable urbanisation process. In Ref. [79], we also proposed protective roadside barriers of a special configuration in the form of discrete shields. Unfortunately, in Ukraine, the use of roadside shields (which are a mandatory attribute of highways in Europe and other countries) is at an early stage. The policy for ensuring a high level of environmental safety of atmospheric air must consider the features of cities and urbanised territories to enhance the effectiveness and efficiency of environmental protection measures and improve the decision-making apparatus at the regional and national levels.

7. Natural and technological safety

Throughout history, natural disasters and anthropogenic catastrophes have claimed lives, inflicted material and financial losses, and created widespread issues across various countries and continents. Until now, no scientific or technical achievements can fully protect people and their places of residence from the consequences of disasters of any genesis. In Ref. [72], the effects of urbanisation processes on the nature of emergencies and the individual risks to the population in different regions have been analysed. We have conditionally divided all regions into four groups depending on the ratio of natural $(Em_n)^n$ and technological hazards $(Em_t)^n$ and index $(I_{edu})^n$. We concluded that the urban impact on the territories is growing, especially in the case of southern Odesa, Kherson, and Mykolaiv regions, which are prone to negative exogenous geological and hydrological processes.

Regions with a sharply increased level of technological accidents and high values of the I_{edu} index (Donetsk, Luhansk, Kyiv, Dnipropetrovsk, Kharkiv, and Zaporizhzhia) are characterised by a powerful industrial and energy complex dominated by extremely dangerous heavy and mining industries and nuclear power. The economic complex of the regions was formed in the context of long-term urbanisation, which was not accompanied by innovation and technological progress, modern re-equipment of production. Outdated (or destroyed on the temporarily uncontrolled territories of Donetsk and Luhansk regions) infrastructure, industry, and energy facilities pose a real threat of large-scale industrial accidents.

Given the fact that the war will undoubtedly lead to a decrease in demographic indicators, let us consider the impact of urban pressure on the regions of the second cluster (described in paragraph 5 of this chapter) with the lowest level of urban load based on environmental safety indicators (**Figure 6**).

Two opposing trends are shown: a decrease in the occurrence of emergencies and an increase in individual risks with the increase in the urbanisation of Ukraine's regions, assessed through the environmental urbanisation indicator. The first trend is demonstrated by the trend line (curve 6 in **Figure 6**) of the disaster hazard of all types $(Em_{tot})^n$, which essentially serves as a summary for other types of hazards. It is important to note its regional characteristics. For instance, the L'viv region is characterised by a relatively high potential for technogenic hazards, which does not correlate with the lowest level of urbanisation in the cluster group. Changes in the placement of infrastructure objects during the war can only increase the region's significance as a major industrial-energy centre and a national-scale transport hub. At the same time, the high level of negative exogenous geological and hydrological processes is not a direct consequence of urbanisation processes and is explained by natural-landscape factors. Overall, the combination of natural and human-induced emergencies synergistically increases the overall hazard potential. The observed trend of decreasing natural-technological emergency indicators with the gradual increase in urbogenic load of regions up to a certain level may indicate the positive consequences of urbanisation. These include a higher level of information and material-technical provision for emergency services, modernisation of technical means, and better organisation of the entire environmental safety management system. However, these positive aspects of urbanisation cannot fully compensate the population risks if the territory of region is prone to various human-induced accidents, natural disasters, and the development of adverse geological processes. Thus, the trend line (curve 5 in **Figure 6**) demonstrates

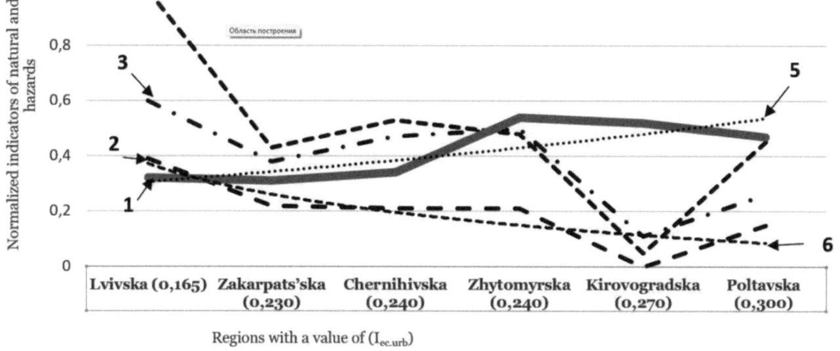

Figure 6.
The feature of natural and technological hazard indicators for the cluster of minimal urban load among the regions of Ukraine: 1—indicator of individual risk of the population (Rind)n; 2—indicator of disaster hazard of all types (Emtot)n; 3—disaster hazard of man-made genesis (Emt)n; 4—indicator of disaster hazard of natural genesis (Emn)n; 5—trend line (Rind)n; 6—trend line (Emtot)n.

a slow increase in individual risks, regardless of the frequency of disasters, with a gradual increase in urban load.

The definitions and correlations of hazard factors and urbanisation indicators conducted have potential for use in relation to Target 11.4: Ensure timely public alert about emergencies through innovative technologies. Differentiation of regions by the levels of individual risks for the population will be useful for determining the algorithm and stages of implementation of this target.

It should be noted that the war in Ukraine creates conditions for increasing the risks of both purely human-induced accidents due to the destruction of techno-sphere facilities and those associated with natural disasters initiated by environmental damage. Therefore, the combination of the factor of environmental urbanisation and the risks of human casualties in disasters can only increase. Therefore, when developing risk-oriented strategies to mitigate the consequences of man-induced accidents and natural disasters at both the national and local levels of management, it is advisable to use indicator assessments of urban features of similar regions that affect the formation of the environmental safety level.

8. Dangers of karst processes

Complex anthropogenic factors primarily have a destructive impact on the geological foundation of urban landscapes, triggering dangerous exogenous geological processes (EGPs) such as landslides, mudflows, karst, flooding, and others. Today, 25% of the world's population lives in areas where karst processes are active [80, 81], which manifest in the form of karst sinkholes or ground collapses. The hazards associated with karst processes are rapidly growing and spreading to urban areas, especially if they were developed without proper planning considering the hydrogeological features of the built-up territories. Cities become centres of degradation of the bedrock of natural landscapes. The destruction of the green cover, the placement of solid and liquid waste, numerous accidents of underground utilities, losses from them of large amounts of water and heat, concentrated mechanical loads on the soil, compaction, and sealing of their surface initiate exogenous processes of the lithosphere group. Karst, flooding, landslides, abrasion, and destruction of river valley slopes cover most of the territory of Ukraine [82] and are activated over significant areas of urban development.

The territorial indicator of karst hazard (R_{karst}) is defined as the proportion of the territory covered by karst rocks of various types in the total area of the region [83]. The coefficient of spatial damage to the territory $(R_{karst})^n$ calculated in this way determines the potential for the maximum possible risk of karst processes within a certain area. The analysis of regional karst formation hazard indicators (R_{karst})n in comparison with the ecological urbanisation indicator $(I_{ec.urb})^n$ shows that in approximately one-third of the regions (Zhytomyr, Kirovohrad, Zakarpattia, Cherkasy, Vinnytsia, Dnipropetrovsk, Kyiv), the increasing levels of karst process hazards almost coincide with the growth of the degree of ecological urbanisation. In these regions, the share of karst landscapes varies from 1.8 to 65.5%, corresponding to a change in the $(R_{karst})^n$ indicator from 0 to 0.645. The correlations between the indicators of geological hazard and urbanisation in the case of low and medium values of karst coverage of regional areas are demonstrated. It is possible to assume a higher probability of urban factors' impact on the frequency of karst manifestations for eight regions—Dnipropetrovs'k, Kyiv, Ivano-Frankivsk, Khmelnitsky, Donetsk, Luhansk,

Kharkiv, and Volyn. For this regional group, high values of the environmental urbanisation indicator $(I_{ec.urb})^n$ (in the range from 0.47 to 1.0) determine the risks of urban initiation of karst manifestations.

The development of karst processes has been studied in dynamics to assess the geological hazards of constantly changing urbanised areas. The consequences of latent processes occurring in the upper layers of surface rocks can be subsidence, collapse, and soil failures both within urban areas and throughout the entire territory of urban impact. For the majority of regions, the area of open karst sinkholes decreased from 19.3 to 11.281 thousand km^2 between 2001 and 2020. Only five regions show the opposite trend (**Figure 7**). The indicator of the increase in the area of open karst $(\Delta S_{k.surf})$ correlates with the indicator of ecological urbanisation of these regions.

The revealed tendency to increase the area of open karst corresponds to the growth of ecological urbanisation due to the exploitation of natural territorial resources, impoverishment of vegetation, and changes in the feature of land use. The assessment of the impact of urbanised areas on the risk of karst processes in the area of hazardous facilities (using the Rivne Nuclear Power Plant (NPP) as an example) is shown [84]. Based on the differentiation of karst threats according to their significance for the safety of the NPP, the most significant types of risks associated with changes in the hydrological regime of soils under the impact of urban areas are proposed. The ways of solving the problem of prevention and reduction of potential hazards of karst process development in the areas of NPPs location are proposed by regulating the risks of urbanised environment impact, limiting urban sprawl in vulnerable areas, and organising an effective system of integrated geo-ecological monitoring.

Urbanisation parameters are directly linked to the vulnerability of karst aquifers to various types of pollution and disruptions in underground drainage. The organisation of surface runoff, the capacity of water supply networks, the purification ability, and the technical condition of sewage facilities are crucial factors that govern the state of karst formations. The execution of engineering-geological works, construction of new techno-sphere objects, and the expansion of areas for economic activities necessitate the application of effective methods for predicting hazardous geological

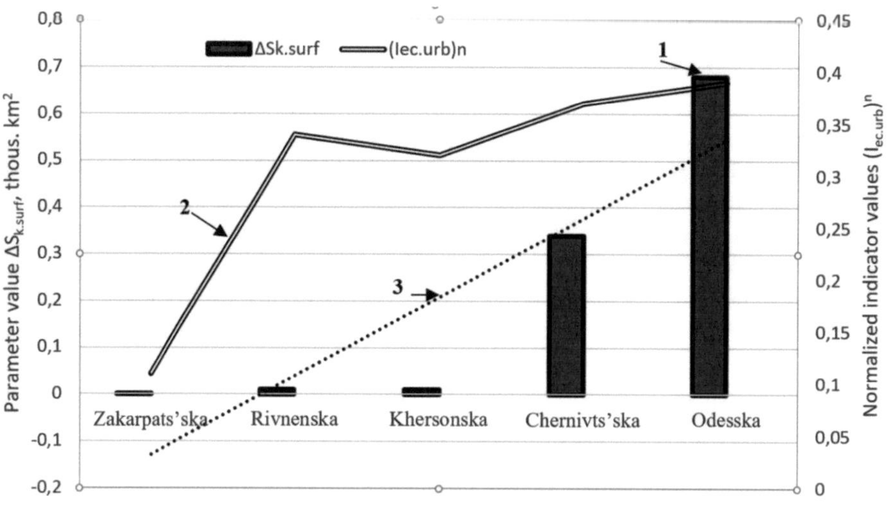

Figure 7.
Expansion of surface karst in the Ukraine regions for the period 2001–2020 in comparison with the ecological urbanisation indicator: 1—indicator $(\Delta S_{k.surf})$*; 2—indicator* $(I_{ec.urb})^n$*; 3—linear trend of the* $(I_{ec.urb})^n$*.*

processes, strict regulation of the level of technological load on karst-prone areas, and establishing boundaries for changes in land use.

9. Conclusions

We have highlighted various facets of the urbanisation process (demographic, industrial or industrial-technological, ecological, agricultural, socio-economic, etc.), which affect the formation of territorial ecological safety both positively and negatively. The ecological aspect acts as a limiting factor for all evolutionary changes due to the scarcity and quality of natural resources. The concept of ecological urbanisation is based on the following key principles:

- The direct and feedback connections between urban environments and the natural systems of adjacent territories develop through a mechanism of positive feedback, which tends to disrupt the balance of the geosystem. This increases the hazard potential of the urban environment and amplifies the adverse outcomes of emergencies.

- The city is considered as an integral object of environmental impact. This eliminates the simplistic procedure of assessing one-way links between urbanisation and the degradation of individual environmental components, even if such impacts are summed up.

It should be emphasised that this chapter does not include a discussion of the compensatory capacity of natural systems to mitigate urbanisation impacts. Indeed, urbanised landscapes are expanding faster than any other type of land use. They are absorbing natural areas that are critical to maintaining ecological balance. SD documents link human well-being to biodiversity, ecosystem conditions, and ecosystem services. The development of green urban infrastructure, the maintenance of green belts, ecological network facilities and ecological corridors within urban agglomerations are essential for sustainable urban development. Indeed, the direction of changing the nature of the ecosystem service of urban ecosystems [71] and the development of ecosystem services of green urban areas [72] have begun to be developed. Recommendations were made to the environmental programmes of Odesa for environmentally oriented urban planning. Unfortunately, this area has been put on hold. The main reason for this is the reorientation of city budgets in times of war. Additionally, fully exploring these aspects is unfeasible without access to remote sensing data and GIS technologies. These restrictions are also related to the war.

The war in Ukraine has significantly altered both the scale of urbanised territories and demographic indicators, particularly in the south-eastern regions where active military actions have devastated dozens of cities. Demographic realities also indicate a population decrease by almost a third, especially in urban areas. However, this has not diminished the value of the indicator as a universal tool for assessing the ecological state of territories.

The proposed methodology of environmental assessments will be essential in determining the recovery strategies for different regions of the country. The integration of sustainable urban planning methods, including green infrastructure, efficient transport systems, and sustainable land use, can significantly affect the state of cities and their surrounding areas. The environmental factor should become a determining one for sustainable and safe development of regions when rebuilding destroyed and new settlements, restoring industrial and infrastructure facilities, and addressing population resettlement issues. It is promising to use the factor of ecological urbanisation to dose the environmental load on territorial resources, in environmental rules of permissible land use, and safe land use. Urban planning parameters must ensure the preservation of natural landscapes and greenery both within and around urban areas, preserve and enhance biodiversity, and provide ecosystem services to city residents.

Acknowledgements

We are grateful to Kiminchidzhi Mariia for her contribution to the collection and processing of statistical data. We also extend our thanks to Butenko O.G., Arsiriy V.A., Surkov S.V., and Smyk S. Yu for their valuable comments.

Conflict of interest

The authors declare no conflict of interest.

Author details

Kateryna Vasiutynska* and Sergey Barbashev
Odesa Polytechnic National University, Odessa, Ukraine

*Address all correspondence to: e.a.vasutinskaya@op.edu.ua

References

[1] The World Bank: Urban Population [Internet]. Available from: https://data.worldbank.org/indicator/SP.URB.TOTL.IN.ZS. [Accessed: December 2, 2023]

[2] State Statistics Service of Ukraine [Internet]. 2023. Available from: https://www.ukrstat.gov.ua/. [Accessed: December 2, 2023]

[3] United Nations Population Division. World Urbanization Prospects: 2018 Revision. [Internet]. Available from: https://www.indexmundi.com/facts/indicators/SP.URB.TOTL.IN.ZS/rankings. [Accessed: December 14, 2023]

[4] OECD Key Environmental Indicators: OECD Environmental Directorate/ Paris, France. 2004. [Internet]. Available from: https://www.oecd.org/environment/indicators-modelling-outlooks/31558547.pdf. [Accessed: December 14, 2023]

[5] OECD Environment Statistics. [Internet]. Available from: https://www.oecd-ilibrary.org/environment/data/oecd-environment-statistics_env-data-en. [Accessed: November 28, 2023]

[6] European Environmental Agency. Indicators. [Internet]. Available from: https://www.eea.europa.eu/en/analysis/indicators. [Accessed: November 28, 2023]

[7] UNECE Working Group on Environmental Monitoring and Assessment: Core Set of Environmental Indicators for Eastern Europe, the Caucasus and Central Asia. [Internet]. Available from: https://unece.org/DAM/env/europe/monitoring/Indicators/Core_indicators_for_EECCA.En.pdf. [Accessed: November 30, 2023]

[8] Sustainable development in the European Union Overview of progress towards the SDGs in an EU context, 2022 edition, Eurostat Supports the SDG s. [Internet]. Available from: https://ec.europa.eu/eurostat/documents/4031688/14665125/KS-06-22-017-EN-N.pdf/8febd4ca-49e4-abd3-23ca-76c48eb4b4e6?t=1653033908879. [Accessed: January 4, 2024]

[9] Liu W, Jiao F, Ren L, Xi X, Wang J, Wang X. Coupling coordination relationship between urbanization and atmospheric environment security in Jinan City. Journal of Cleaner Production. 2018;**204**:1-11. DOI: 10.1016/j.jclepro.2018.08.244

[10] Liang K, Wang Z, Ji L. The effect of urbanization on environmental pollution in rapidly developing urban agglomerations. Journal of Cleaner Production. 2019;**237**:117649. DOI: 10.1016/j.jclepro.2019.117649

[11] Zhang Z, Xia F, Yang D, Chen Y. Discussion of an environmental depletion assessment method – A case study in Xinjiang, China. PLoS One. 2022;**17**(1):e0262092. DOI: 10.1371/journal.pone.0262092

[12] Yang J, Ren J, Sun D, Xi X, Xia J-C, Jin C, et al. Understanding land surface temperature impact factors based on local climate zones. Sustainable Cities and Society. 2021;**69**:102818. DOI: 10.1016/j.scs.2021.102818

[13] Yu D, Shi P, Liu Y, Xun B. Detecting land use-water quality relationships from the viewpoint of ecological restoration in an urban area. Ecological Engineering. 2013;**53**:205-216. DOI: 10.1016/j.ecoleng.2012.12.045

[14] Junguo L, Qingying L, Hong Y. Assessing water scarcity by

simultaneously considering environmental flow requirements, water quantity, and water quality. Ecological Indicators. 2016;**60**:434-441. DOI: 10.1016/j.ecolind.2015.07.019

[15] Carey RO, Migliaccio KW, Li Y, Schaffer B, Kiker GA, Brown MT. Land use disturbance indicators and water quality variability in the Biscayne Bay watershed, Florida. Ecological Indicators. 2011;**11**(5):1093-1104. DOI: 10.1016/j.ecolind.2010.12.009

[16] Arfanuzzaman M, Atiq Rahman A. Sustainable water demand management in the face of rapid urbanization and ground water depletion for social–ecological resilience building. Global Ecology and Conservation. 2017;**10**:9-22. DOI: 10.1016/j.gecco.2017.01.005

[17] Pires A, Morato J, Peixoto H, Botero V, Zuluaga L, Figueroa A. Sustainability assessment of indicators for integrated water resources management. Science of The Total Environment. 2017;**578**:139-147. DOI: 10.1016/j.scitotenv.2016.10.217

[18] Burkhard B, Müller F. Drivers-pressure-state-impact-response. Encyclopedia of Ecology. 2008;**5**:967-970. DOI: 10.1016/B978-008045405-4.00129-4

[19] Brambila A, Flombaum P. Comparison of environmental indicator sets using a unified indicator classification framework. Ecological Indicators. 2017;**83**:96-102. DOI: 10.1016/j.ecolind.2017.07.023

[20] Development. Group on Environmental Performance, and Development. Group on the State of the Environment. OECD core set of indicators for environmental performance reviews. In: Enviromonographs №. 83. Organisation for Economic Co-Operation and Development. Paris: OECD; 1993

[21] Maurya SP, Singh PK, Ohri A, Singh R. Identification of indicators for sustainable urban water development planning. Ecological Indicators. 2020;**108**:105691. DOI: 10.1016/j.ecolind.2019.105691

[22] Ya L, Yi L, Zhou Y, Shi Y, Zhu X. Investigation of a coupling model of coordination between urbanization and the environment. Journal of Environmental Management. 2012;**98**:127-133. DOI: 10.1016/j.jenvman.2011.12.025

[23] Ozturk I, Al-Mulali U. Investigating the validity of the environmental Kuznets curve hypothesis in Cambodia. Ecological Indicators. 2015;**57**:324-330. DOI: 10.1016/j.ecolind.2015.05.018

[24] Wang S-J, Ma H, Zhao Y-B. Exploring the relationship between urbanization and the eco-environment—A case study of Beijing–Tianjin–Hebei region. Ecological Indicators. 2014;**45**:171-183. DOI: 10.1016/j.ecolind.2014.04.006

[25] Vavrova K. Environmental indicators. In: Proceedings of the Conference on Current Problems of the Corporate Sector, 2020. SHS Web of Conferences. Vol. 83. France: EDP Sciences; 2020. p. 01070. DOI: 10.1051/shsconf/20208301070

[26] Ma L, Xu W, Zhang W, Ma Y. Effect and mechanism of environmental regulation improving the urban land use eco-efficiency: Evidence from China. Ecological Indicators. 2024;**159**:111602. DOI: 10.1016/j.ecolind.2024.111602

[27] Gonzalez S, Belen M, Ortega E, Azevedo JC. Relationships between amount and configuration indicators in real-world landscapes: Novel references for the assessment of forest cover in the

Iberian Peninsula. Ecological Indicators. 2024;**159**:111634. DOI: 10.1016/j.ecolind.2024.111634

[28] Pukowiec-Kurda K. The urban ecosystem services index as a new indicator for sustainable urban planning and human well-being in cities. Ecological Indicators. 2022;**144**:109532. DOI: 10.1016/j.ecolind.2022.109532

[29] Sachs JD, Kroll CH, Lafortune G, Fuller G, Woelm F, editors. The Decade of Action for the Sustainable Development Goals: Sustainable Development Report 2021. Cambridge: Cambridge University Press; 2021. 518 p. DOI: 10.1017/9781009106559

[30] The Law of Ukraine "On the Main Principles (Strategy) of the State Environmental Policy of Ukraine for the Period up to 2030" of 28.02.2019 No. 2697-VIII / Verkhovna Rada of Ukraine. Available from: https://zakon.rada.gov.ua/laws/show/2697-19. [Accessed: December 22, 2023]

[31] Sustainable Development Goals Ukraine. 2019. Monitoring Report. [Internet]. Available from: https://www.ukrstat.gov.ua/csr_prezent/2020/ukr/st_rozv/publ/SDGs13.01.2020_engl.pdf. [Accessed: December 22, 2023]

[32] Sustainable Development Goals Ukraine. 2020. Monitoring Report. [Internet]. Available from: https://www.ukrstat.gov.ua/csr_prezent/2020/ukr/st_rozv/publ/SDGs_Ukr_2020_Monitoring_Report_eng.pdf. [Accessed: December 22, 2023]

[33] Sustainable Development Goals Ukraine. 2021. Monitoring Report. [Internet]. Available from: https://www.ukrstat.gov.ua/csr_prezent/2020/ukr/st_rozv/publ/SDGs%20Ukraine%202021%20Monitoring%20Report%20engl.pdf. [Accessed: December 22, 2023]

[34] Redko K, Miroshnychenko V. Research on sustainable development in Ukraine: Assessment of goal achievement. Entrepreneurship and Innovation. 2022;**22**:5-13. DOI: 10.37320/2415-3583/22.1. [Accessed: December 23, 2023]

[35] Sergienko LV. Assessment of sustainable development indicators of urbanization areas and cities of Ukraine. Public Management and Administration. 2022;**28**:207-214. DOI: 10.32782/pma2663-5240-2022.28.41. [Accessed: December 23, 2023]

[36] Klymenko MO, Pryshchepa AM, Brezhytska OA. Assessment of the City Territories' Condition Based on Sustainable Development Indicators: Monograph. Rivne: NUWGP; 2018. 221 c

[37] Kyryn RS. Legal tools for implementing strategic directions for the development of the City's environmental security system: International and European levels. Economics and Law. 2022;**2**:39-55. DOI: 10.15407/econlaw.2022.02.039

[38] Stolberg FV, Kovalenko YL. Improvement of the environmental monitoring system and indicators of the environmental dimension within the framework of the sustainable development strategy in Ukraine. Municipal economy of cities. 2016;**132**:88-92

[39] Grechanyk NY. Ecological dimension of sustainable development. A marketing approach. Sustainable development of the economy. 2014;**1**(23):94-101

[40] Zgurovskyi YV. Ukraine in the Indicators of Sustainable Development. K: NTU "KPI"; 2014. 172 c

[41] Lelechenko AP. Ensuring environmental safety of a large city in the context of sustainable

development. Investments: practice and experience. 2022;**5—6**:38-42. DOI: 10.32702/2306-6814.2022.5—6.38

[42] Skok SV, Stratichuk NV. Scientific and methodological aspects of assessing the sustainable development of urban ecosystems. Ecological sciences. 2020;**1**(28):367-372. DOI: 10.32846/2306-9716/2020.eco.1-28.58

[43] Ivaniuta SP. On the integrated assessment of environmental safety of the regions of Ukraine. Ecological safety and nature management. 2013;**13**:34-44. [Internet]. Available from: http://nbuv.gov.ua/UJRN/ebpk_2013_13_6. [Accessed: December 24, 2023]

[44] Varlamov EM, Tolstykh YO. Analysis of the formation and use of ecological and economic indicators and indicators for assessing sustainable development of the region. Information processing systems. 2011;**3**(93):165-168

[45] Tyutyunyk VV, Ivanets HV, Horylyshev SA. Methodology for assessing the level of technogenic, natural and social hazard of administrative and territorial units of Ukraine. Collection of Scientific Works of the National Academy of the National Guard of Ukraine. 2016;**1**(27):29-37

[46] Kolesnik VY, Borysovs'ka OO, Pavlychenko AV, Shirin AL. Determination of the trends and regularities of occurrence of emergency situations of technogenic and natural character in Ukraine. Науковий вісник НГУ. 2017;**6**:124-131

[47] Ivanets G, Tolkunov I. Complex model for predicting losses due to emergencies in the state. Control, navigation and communication systems. Collection of scientific papers. 2018;**6**(52):68-73. DOI: 10.26906/SUNZ.2018.6.068

[48] Vasutynska KA, Barbashev SV. The analysis of the principles and methods evaluation of environmental safety levels in regional context. Proceedings of Odessa Polytechnic University. 2017;**3**(53):114-121

[49] Geographical Encyclopaedia of Ukraine, Volume 3 (P to Z). Kyiv: M.P. Bazhan Ukrainian Encyclopaedia

[50] Voloshyna YY. Theoretical Foundations of the Concept of Suburban Area and its Main Interrelations with the City. Vol. 10. Available from: http://www.economy.nayka.com.ua/pdf/10_2015/45.pdf. Effective economy; 2015. [Accessed: December 24, 2023]

[51] Loibl W, Etminan G, Gebetsroither-Geringer E, Neumann HM, Sanchez-Guzman S. Characteristics of urban agglomerations in different continents: History, patterns, dynamics, drivers and trends [Internet]. In: Urban Agglomeration. InTech; 2018. DOI: 10.5772/intechopen.73524

[52] On approval of the State Strategy for Regional Development for 2021-2027: Resolution of the Cabinet of Ministers of Ukraine of 05 August. 2020 p. 695. Available from: https://www.kmu.gov.ua/npas/pro-zatverdzhennyaderzhavnoyi-strategiyi-regionalnogo-rozvitku-na-20212027-t50820. [Accessed: December 24, 2023]

[53] OECD. Redefining "Urban": A New Way to Measure Metropolitan Areas. Paris: OECD Publishing; 2012. DOI: 10.1787/9789264174108-en

[54] Draft law from 17.07.2017 №6743. Available from: http://w1.c1.rada.gov.ua/pls/zweb2/webproc4_1?pf3511=62318. [Accessed: December 24, 2023]

[55] Draft law from 19.12.2019 № 2637. Available from: http://w1.c1.rada.gov.ua/

pls/zweb2/webproc4_1?pf3511=67710. [Accessed: December 24, 2023]

[56] Council of Europe Office in Ukraine. Available from: https://www.coe.int/en/web/kyiv/-/metropolitan-governance-in-ukraine

[57] Steshenko TV. Urban agglomerations as a form of cooperation of territorial communities legal scientific. Electronic Journal. 2023;**3**:114-117. DOI: 10.32782/2524-0374/2023-3/25

[58] Stepanenko AV, Omelchenko AA. Urban agglomerations as a form of the modern world urbanisation process. State and Regions. Series Economics and Entrepreneurship. 2019;**3**(108):184-192

[59] Vasutynska KA, Barbashev SV, Butenko OG, Surkov SV. Determination of the urbans indicator as a complex index of the environmental safety of city course territory system. In: Collection of Scientific Papers of the XIII International Scientific and Technical Conference "Problems of Environmental Safety". Ukraine, Kremenchuk: Kremenchuk National University; 2-4 October 2019. pp. 212-215

[60] Ministry of Environmental Protection and Natural Resources of Ukraine. Environmental Passports of the Regions [Internet]. Available from: https://mepr.gov.ua/diyalnist/napryamky/ekologichnyj-monitoryng/ekologichni-pasporty/. [Accessed: January 13, 2024]

[61] Vasiutynska K, Barbashev S, Kiminchydzhy M. The risk of a water scarcity in the Ukraine regions in urbanization terms. Ecological Science. 2020;**4**(31):42-48. DOI: 10.32846/2306-9716/2020.eco.4-31.6

[62] Vasiutynska K. Assessment of urbogenic load on territory and population of regionally important Ukraine cities by emissions from stationary sources of air pollution. Ecological Science. 2021;**4**(37):102-108. DOI: 10.32846/2306-9716/2021.eco.4-37.15

[63] Vasutynska K, Barbashev S. Analysis of dynamics of man-made fires in conditions of urbanization in Ukraine. Technology Audit and Production Reserves. 2018;**4**(3(42)):16-23. DOI: 10.15587/2312-8372.2018.141376

[64] Vasutynska K. Assignment of the new type of ecological services for providing human safety under conditions of urban environment. EUREKA: Life Sciences. 2018;**2**:9-18. DOI: 10.21303/2504-5695.2018.00598

[65] Vasiutynska K. Assessment of the ecosystem service indicators of urban green zones in relation with the urbogenic load of Ukraine regions. Ecological Science. 2021;**1**(34):36-43. DOI: 10.32846/2306-9716/2021.eco.7-34.7

[66] European Environment Agency. EEA's Conceptual Framework for Urban Sustainability. Modified 14 Apr 2023. Available from: https://www.eea.europa.eu/en/topics/in-depth/urban-sustainability. [Accessed: December 27, 2023]

[67] Vasiutynska K, Barbashev S, Kiminchydzhy M. Evaluation of the environmental urbanization's complex indicator of the Ukraine regions. Ecological Science. 2020;**3**(30):7-14. DOI: 10.32846/2306-9716/2020.eco.3-30.1

[68] Komarnitskaya GO, Shipulina YS, Ilyashenko NS. Influence of urbanisation on the innovative development of the regions of Ukraine. Marketing and management of innovations. 2017;**3**:336-345. Available from: http://mmi.fem.sumdu.edu.ua/. [Accessed: January 10, 2024]

[69] Shevchenko O, Vlasiuk O, Stavchuk I, Vakoliuk M, Ilyash O. Assessment of Vulnerability to Climate Change. Ukraine: The Eastern Partnership Climate Forum (EPCF) and the Working Group of Civil Society Organisations on Climate Change (WG CSO CC); 2014. Available from: https://necu.org.ua/wp-content/uploads/ukraine_cc_vulnerability.pdf. [Accessed: January 10, 2024]

[70] Jupyter Notebook [Internet]. Available from: https://jupyter.org/

[71] Gavryliuk OK. Analysis of approaches to assessing the level of urbanisation: The case of the northeastern macro-region of Ukraine scientific bulletin of Kherson state university. Series Geographical Sciences. 2018;**8**:21-31

[72] Vasiutynska K., Barbashev S. Indicator assessment of the impact of urbanization processes on the state of natural and man-caused safety in the Ukraine regions. In: Sustainable Development: Environmental Protection. Energy Saving. Balanced Nature Management". Lviv: ZUKC LLC; 2020. p. 232-255. DOI: 10.23939/book.ecocongress.2020

[73] WHO's First Global Conference on Air Pollution and Health Improving air quality, Combatting Climate Change – Saving Lives. Available from: https://www.who.int/airpollution/events/conference/WHO_Conference_on_Air_Pollution_and_Health-Concept_Note-final.pdf. [Accessed: December 27, 2023]

[74] Kinney PL. Interactions of climate change, air pollution, and human health. Current Environmental Health Reports. 2018;**5**:179-186. DOI: 10.1007/s40572-018-0188-x/

[75] Baklanov A, Molina LT, Gauss M. Megacities, air quality and climate. Atmospheric Environment. 2016;**126**:235-249. DOI: 10.1016/j.atmosenv.2015.11.059

[76] Numbeo.com. Europe: Current Pollution Index de City. Available from: https://www.numbeo.com/pollution/region_rankings_current.jsp?region=150. [Accessed: January 10, 2024]

[77] World Air Quality Report 2020: Region and City PM2.5 Ranking. Available from: https://www.iqair.com/world-air-quality-report (дата звернення 4.08.2021)

[78] Vasiutynska K, Barbashev S. Impact assessment of the urbanization factors on the atmosphere pollution in the Ukrainian regions. Bulletin of Kremenchuk Mykhailo Ostrohradskyi National University. 2021;**4**(129):83-90. DOI: 10.30929/1995-0519.2021.4.83-89

[79] Arsiriy VA, Vassyutinskaya EA, Smyk SY, V.A. Chumachenko optimization of air flow while using roadside protective Shildings in the urban highway system. Ecology and industry of Russia. 2022;**2**(26):26-31. DOI: 10.18412/1816-0395-2022-2-26-31

[80] Ferreira CSS, Walsh RPD, Ferreira AJD. Degradation in urban areas. Current Opinion in Environmental Science & Health. 2018;**5**:19-25. DOI: 10.1016/j.coesh.2018.04.001

[81] Guttierrez F, Parise M, De Waele J, Jourde H. A review on natural and human-induced geohazards and impacts in karst. Earth-science Reviews. 2014;**138**:61-88. DOI: 10.1016/j.earscirev.2014.08.002

[82] DNVP "Geoinform Ukrainy": Information Yearbook on the Activation of Hazardous Exogenous Geological Processes in the Territory of Ukraine Based on Monitoring Data EGP. Available

from: https://geoinf.kiev.ua/publikatsiyi/shchorichnyky/shchorichnyk-egp/

[83] Vasiutynska K, Barbashev S. The impact assessment of the urbanization factors on the hazard of the expansion of karst rocks in the Ukraine regions. Geochemistry of Technogenesis. 2021;**5**:33-40. DOI: 10.15407/geotech2021.33.034

[84] Vasiutynska K, Barbashev S, Smyk S. Impact of urban land on the risk of karst processes in the NPP location area (on the example of Rivne NPP). Problems of Atomic Science and Technology. 2023;**3**(145):120-126. DOI: 10.46813/2023-145-120

Chapter 5

Changes of Essential Facilities in Housing Estates in an Aging Society: The Failure of City Planning in Japan

Yoshimichi Yui and Tomoko Kubo

Abstract

Shrinking suburbs cause to occur several problems in Japan. One of the serious problems is the safety of communities, which are kept by community's members. Some suspicious persons will invade vacancy houses, because nobody can maintain them. And it is difficult to keep house, because of aging of residents. And many public facilities and shopping centers were closed by the depopulation. This chapter introduces two case studies. One is the case of suburb of Hiroshima and another is Ryugasaki new town which is located in suburb of Tokyo. Amidst the aging in the suburb of Hiroshima, facilities for the elderly have been increasing in suburban housing estates in recent years. Facilities planned for child-rearing households and facilities developed for those households are no longer consistent with aging population attributes, and improvement of living convenience facilities and welfare facilities for the elderly is required. Therefore, this study focuses on the location of nursing care insurance service projects such as nursing home care and home care facilities, among welfare facilities for aged people in suburban housing estates, and examines the location of nursing care welfare facilities in the suburban housing estates of Hiroshima City. And we will point out the failure of Japanese town planning.

Keywords: aging society, housing estates, town planning, essential facilities, welfare facilities, Japanese cities

1. Introduction

The suburbs are not a dream location for every Japanese family, and depopulation has occurred in several urban regions. In many traditional urban growth models, suburbs continuously grow, with population inflow to commuter towns in metropolitan regions. However, this trend is changing, due to the population decreasing in old suburban areas, while the population in inner-city areas increases. Kubo and Yu [1] point that old Japanese suburban housing estates, which were developed before the 1970s, are in decline, and they have encountered several serious problems. The most serious are aging residents and a decreasing population, which are caused by long-term dwellings. As many Japanese think that the "Japanese Dream" is the occupancy of a detached house in a suburb, suburban Japanese residents tend to stay put after childrearing. Another serious problem is the increase in the number of vacant houses [2]. In this study, we attempted to clarify the conditions of shrinking suburbs in Japanese cities and introduce revitalization activities in the suburbs, with the aim of contributing to the revitalization of suburbs through geographical studies.

In Japan, there is little poverty and diversification in the suburbs. Aging, however, is a serious problem. Aging has resulted in an increase in the number of vacant houses and shrinking communities [1]. Regarding the aging of suburban residents in old housing, the first generation of migrants grew older and continued to live in their own houses in suburban areas. Furthermore, their children grew up and moved out. Therefore, aging communities without younger generations have formed in the suburbs. These were caused by the failure of town planning, which supplied the same types of houses for the short term [3]. Furthermore, increases in vacant houses are seen throughout old suburban housing estates, which induce new uneasiness, social troubles, and a drop in house prices. As a countermeasure, some suburban communities have attempted to vitalize and promote community activities.

Rapid population aging in modern urban societies has brought about significant changes in various aspects of social systems. In most metropolitan regions in Japan, suburban housing estates have aged rapidly in recent years. This is because, in the high economic growth period, a large number of monotonous houses without diversity in the layout and sales price zone were supplied in large quantities, and the life stage, age composition, and social position were almost homogeneous [1]. For this reason, in many suburban housing estates, the number of older persons increased, owing to the loss of the young generation (relative aging), and the total number of older persons increased, owing to the aging of residents (absolute aging). As these progressed in parallel, the aging state progressed remarkably.

The shrinking suburbs can cause several problems. One serious problem is the safety of communities, which is maintained by community members. Some people invade vacant houses because no one can maintain them. It is difficult to maintain houses because of aging residents. It is difficult to keep houses and gardens clean. Many public facilities and shopping centers have been closed due to depopulation. People must shop by bus or train. Although older adults depend on cars, they can no longer drive independently [3].

The suburban housing estates often assume the child-rearing generation as a resident at the beginning of development, and the facilities in the residential estates are educational facilities, such as kindergartens, child-rearing facilities, and elementary schools. Many commercial facilities are arranged systematically. However, owing to the aging of the residents, the number of children decreases significantly in elementary schools, empty classrooms occur, and commercial facilities with products for young parenting households in the center of housing estates are needed to respond to the needs of the old persons. There are also some places where shops close sequentially, and shutters are in place [1].

Amidst this aging trend, facilities for the old persons have been increasing in suburban housing estates in recent years. Facilities planned and developed for child-rearing households are no longer consistent with aging population attributes, and improvements in living convenience and welfare facilities for the old persons are required.

This study aimed to clarify the changes in suburban neighborhoods in Japanese metropolitan regions. In particular, we focused on aging and changes in family members to clarify the shrinking conditions of the suburbs. The second aim was to discuss how town planning caused these crises and how Japanese housing choices were made.

In the following sections, we present the results of an analysis of aging and regional transformation in suburban areas using two case studies: Ryugasaki City in the Tokyo metropolitan area and Hiroshima City in the regional metropolitan area.

2. Case study of Ryugasaki New Town

2.1 Development of Tokyo's outer suburbs

After World War II, the Japanese government established national land-planning regimes in the 1950s. First, a hierarchical decision-making structure for planning (national, regional, and prefectural land planning) was developed to reconstruct urban infrastructure and overcome housing shortages. In the 1960s, a series of national land-planning and related policies were designed to stimulate rapid industrialization and urbanization, not only in the three major metropolitan areas but also in newly designated industrial areas, to form industrial clusters throughout Japan. At the same time, transportation systems (e.g., highway networks, bullet trains, nationwide train systems, and airport networks) were intensively established, resulting in the growth of major large-city networks connected by management functions (e.g., the network between head offices in Tokyo, Osaka, and Nagoya and their regional offices in Sapporo, Sendai, Hiroshima, Okayama, and Fukuoka). Under these circumstances, the rapid inflow of younger workers into major cities required new frontiers to accommodate them. As a result, suburban development has increased since 1980, peaking in the 1980s, just before Japan's economic bubble burst.

In terms of housing development to overcome housing shortages and accommodate newly migrated workers to large cities, the Japan Housing Corporation (presently the Urban Renaissance Agency) played a significant role. To supply a sufficient number and quality of housing in a short period, the Japanese government launched the Housing Loan Corporation (1950), enacted the Public Housing Law (1951), and established the Japan Housing Corporation (1955). The Housing Loan Corporation supported housing purchases by high-income populations and the Public Housing Law supported young couples and nuclear families in offering affordable rental housing units. The Japan Housing Corporation was responsible for huge suburban housing developments; usually, these included detached houses, rental housing units, and shopping streets, and many of them learned from the planning ideas of neighborhood units. In one catchment area of an elementary school, a neighborhood unit, community center(s), shopping and office areas, and a variety of housing were developed. Each facility and housing area was connected by pedestrian roads separated from automobile roads. In Osaka, Senri New Town (NT) has welcomed new residents since 1962, followed by Kozoji NT in Nagoya in 1968 and Tama and Narita NTs in Tokyo in 1971. Similar suburban developments were popular between the 1960s and the early 1990s. In addition to public agents (Japan Housing Corporation and prefectural housing corporations), private housing developers joined the suburban housing supply, reaching a home-ownership rate of 60% in Japan in the 1960s [4].

2.2 Case study of Ryugasaki New Town

Using the case study of Ryugasaki NT, which has been developed by the Former Japan Housing Corporation since 1981 at the edge of the Tokyo metropolitan area, we explored the possibilities for suburban neighborhoods to be sustainable residential areas.

In 1981, the western half of NT was completed and occupied by thousands of residents commuting to central Tokyo, whereas the eastern half was under construction. After four decades, the western half of NT is now occupied by elderly couples or singles, turning them into super-aging neighborhoods, along with the outmigration of the younger generation (**Figure 1**).

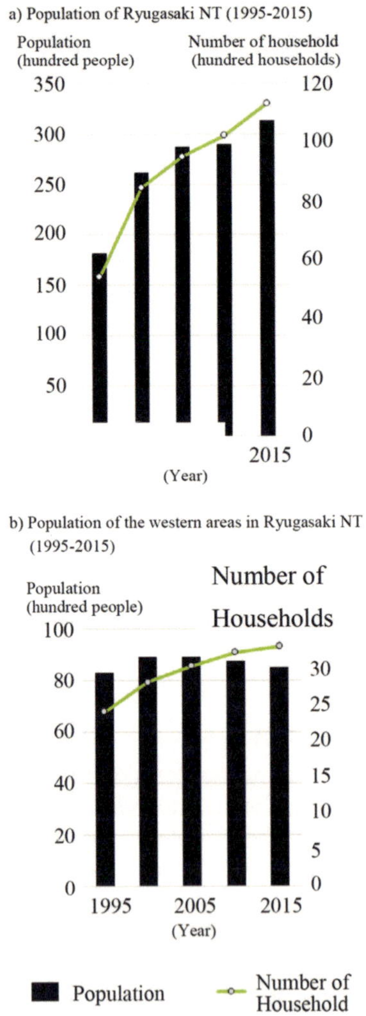

Figure 1.
Changes in population and households in Ryugasaki NT (1995–2015). Source: Population Census of Japan (1995–2015).

According to our field survey, slight but constant inflows of foreigners, renters, and buyers of existing houses contributed to maintaining the population in the western half of the NT. However, the eastern parts of NT took more than 15 years to complete until the municipal government changed its main area from residential or industrial to commercial in the late 1990s. After the opening of a new shopping center in the mid-1990s, the surrounding residential areas were purchased by relatively younger families who worked and studied in neighboring cities. Ryugasaki NT complemented the residential function of the Tokyo metropolitan area in the 1980s and the 1990s, but gradually weakened this function. Now, NT supplements the strong housing demand of local, mainly blue-collar residents. Such a transition in the major functions of NT helped make an aging suburban housing estate a sustainable residential area. The details are presented in the following sections.

2.3 Changes in the residential characteristics of Ryugasaki NT

We conducted questionnaire surveys in Ryugasaki NT in May 2021 and distributed them to all the households in the NT on behalf of all community centers there; 11,719 questionnaires were distributed, and 2130 valid answers were collected (18.2% of respondents). We compared valid answers between the eastern and western areas.

Figure 2 shows the locations of the major workplaces of household heads in the two areas of the NT. In the western area, most household heads worked in central Tokyo, whereas Ryugasaki and its neighboring cities became remarkable in the eastern parts of the NT. In our questionnaire surveys, a large proportion of Western residents searched for their homes when they married, had babies, or felt a shortage of housing space because of the growth of their children in a large part of the Tokyo metropolitan area. Due to rising house prices at the time, they had to give up on the idea of purchasing housing in the near suburbs (10–30 km commuter belt of Tokyo).

Meanwhile, with a decline in house prices in the 1990s and the first decade of the twenty-first century, residents of the eastern part of the NT could find affordable housing options in Ryugasaki NT. At that time, the former Japan Housing Corporation stopped new housing developments and was one of the rare developers that offered total planning for the NT area. They had plans for built environments and facilities to satisfy residents' daily needs (e.g., shopping, recreation, education, communication among residents, or medical treatment). Private developers tend to sell only houses and land. Therefore, most residents tended to live and work in neighboring areas and evaluated the residential environment of Ryugasaki NT.

The residential areas of adult children clearly reflected the differences between the two areas (**Figure 3**). In the western part of the area (top figure), a large proportion of adult children lived in Tokyo, followed by those living in neighboring areas of their parents. In addition, many adult children maintained a double-income status and worked in offices in central Tokyo or nearby suburbs. By contrast, in the eastern areas, most adult children lived nearby and maintained a double-income status.

In summary, the western area is characterized by typical suburban residents from the 1970s to the 1980s, mostly single-income households working in central Tokyo supported by housewives. However, the eastern area is characterized as a great housing option for residents who live and work in neighboring areas. Ryugasaki NT is part of the Tokyo metropolitan area, compensating for residential functions during the rapid growth period from the 1980s to the 1990s. In the period of the urban divide and shrinking suburbs in the first two decades of the twenty-first century [5],

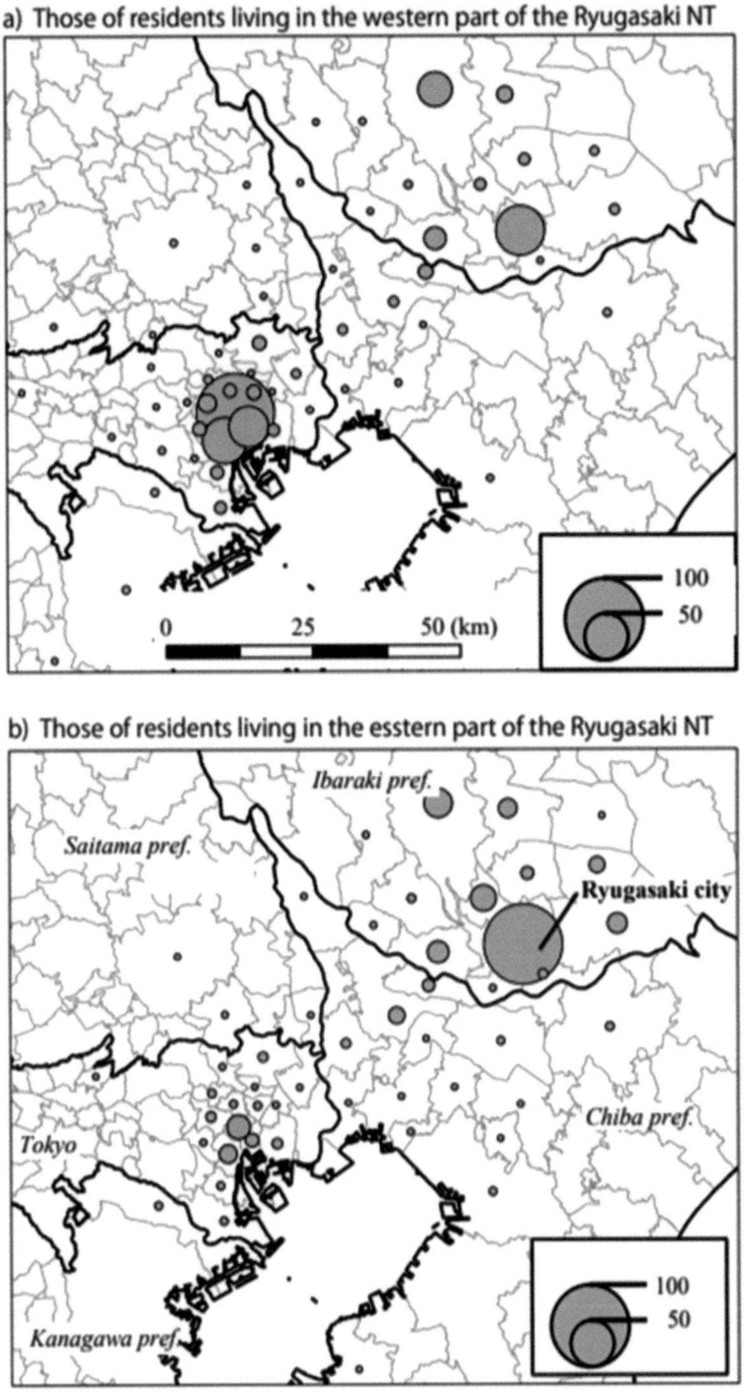

Figure 2.
The location of the offices to which household heads worked for longest. Note: In the western part of the NT (top figure), out of 567 answers from residents of this area, 309 were listed on the map. In the eastern part of the NT (bottom figure), of the 232 answers from residents of this area, 217 were listed on the map. Locations outside of the map were excluded. Source: Authors' questionnaire surveys.

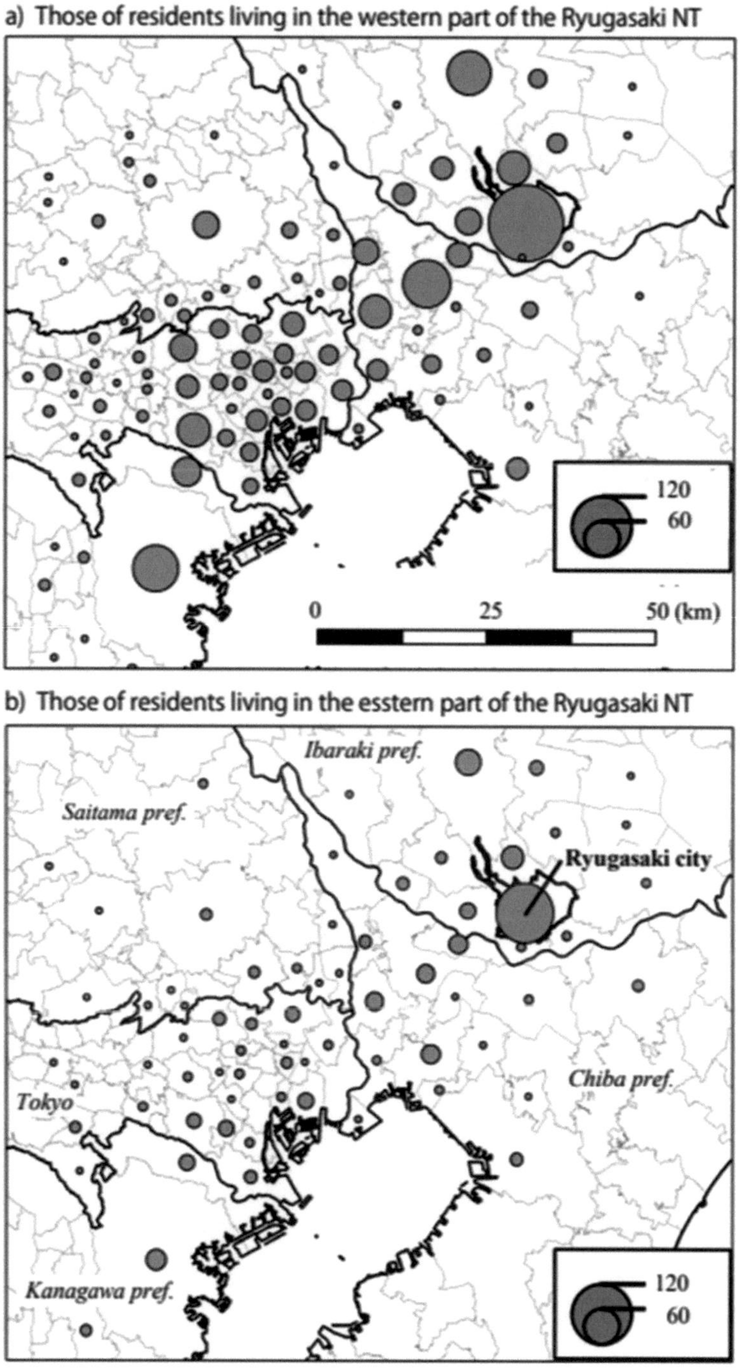

Figure 3.
The residential areas of adult children. Note: In the western part of the NT (top figure), out of data from 1034 adult children residents of this area, 861 were listed on the map. In the eastern part of the NT (bottom figure), among the 361 adult children residing in this area, 330 were listed on the map. Locations outside the map and adult children living with their parents were excluded. Source: Authors' questionnaire surveys.

Ryugasaki NT played the role of a center for neighboring cities, offering superior housing options to other small and partial housing developments.

3. Constant inflow of residents to avoid aging and shrinking neighborhoods in Japan

We have confirmed that western and eastern areas within Ryugasaki NT had different urban functions, reflecting the different economic and urban conditions between 1970 and the 1990s and the first two decades of the twenty-first century. In the Japanese housing market, most of the shrinkage in the outer suburbs was triggered by the aging of the whole neighborhood due to the aging of existing residents and outflows of younger generations under the fragile existing housing market [6]. Therefore, we examined the inflow of new residents in a district in the western area (**Figure 4**).

According to the interview surveys, a significant number of residents moved out when the "Building Agreements" of this neighborhood were not renewed. In Ryugasaki NT, most of the neighborhoods established "Building Agreements" to maintain pleasant residential environments, but some neighborhoods forgot to renew, gave up renewing due to the lack of agreement from residents, or modified the contents to meet today's needs.

Figure 4.
The time period when residents moved into a neighborhood in the western part of Ryugasaki NT (1991–2018). Note: We determined the time period for residents moving into the residence using name plates written on residential maps. If name plates were renewed between the two time periods, we judged that the resident moved between that time period. Source: ZENRIN Co. Ltd., Residential Maps, Ryugasaki City (1991/2000/2009/2018).

In this case neighborhood, the renewal of "Building Agreements" was not successful and caused an outflow of residents. The vacant housing was occupied by new residents, and the constant and slight inflows of new residents have continued today. Although the neighborhood is aging, these constant inflows prevent complete shrinkage.

To calculate the quality of the residential environments of the case neighborhood in the western area of the Ryugsaki NT, we conducted a field survey to determine the condition of each house and its site based on two criteria, "Building Agreements" and contemporary and universal design requirements. Details of the calculation and content are explained in the notes for each figure. There was no significant correlation between these two criteria.

According to the score distribution based on "Building Agreements," only southern blocks showed a high score (**Figure 5**). These blocks are occupied by

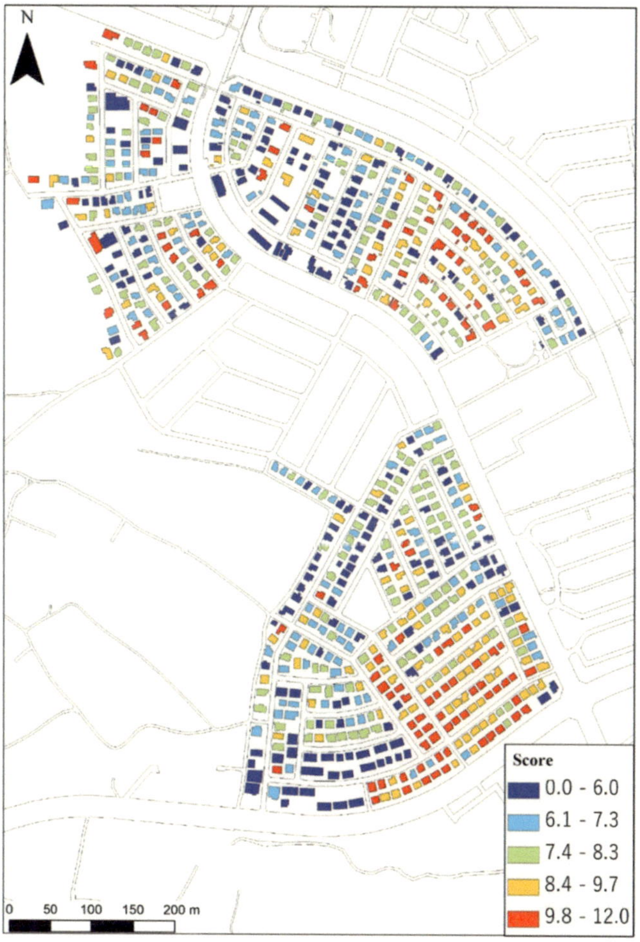

Figure 5.
The score distribution based on the "building agreements" of the case neighborhood (2021). Note: The "Building Agreements" of this case neighborhood were considered for (1) land use (maximum 4 points: Detached house (4) and others (0)), (2) trees and planting (maximum 4 points: Based on the number of trees in the garden), and (3) external structure (maximum 4 points: Tree material (2), other plants (1), others (0); height is 150 cm or lower (2) and over 150 cm (0)). Source: Authors' field surveys conducted in March 2021.

education-minded volunteer residents (according to interview surveys). In addition, these blocks face the elementary school across the road, therefore many houses had security signs (e.g., signboards, stickers, or flags), such as "don't rush into the road" or "be careful of crime."

With regard to more contemporary needs (more than two parking spaces) and universal design guidelines (no barriers between the road and entrance of the house), the northern blocks showed high scores (**Figure 6**). Essentially, the blocks with a high score in the "Building Agreements" criteria received a lower score here. High-scoring blocks are occupied by newly moved residents or renovated houses; therefore, they

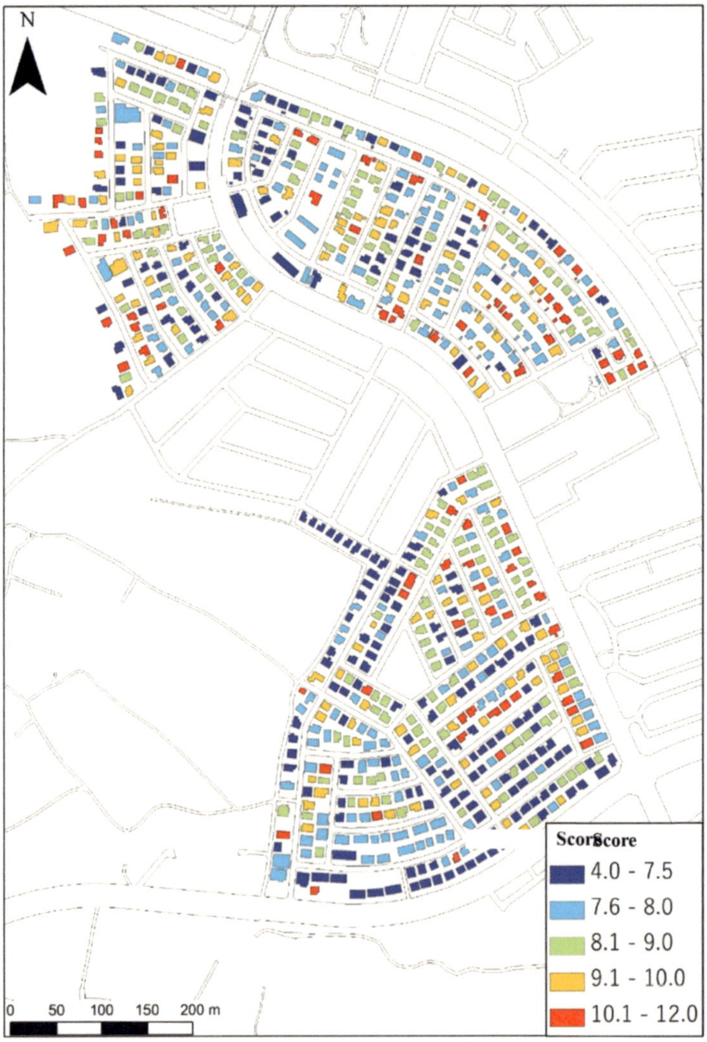

Figure 6.
*The score distribution based on requirements for universal design (2021). Note: The score was calculated using the following: (1) No barrier between the road and the housing site (maximum 4 points) and the site to the entrance of the house (4), based on the formula of "4-(number of steps)/2," if the number of steps exceed 8, the score is 0, but we add 2 points if there are handrails for overcoming the barrier; and (2) the number of parking spaces (formula is "the number of parking spaces*2, with a maximum 4 of points). Source: Authors' field surveys conducted in March 2021.*

have sufficient parking spaces and fewer barriers between the road and entrance. Due to the fragile existing housing market of Japan, new ideas, such as "barrier free" environments, can be seen regularly in newly built houses but not in existing houses. Homeowners who require new facilities must continue investing in their homes to meet their residential needs. The fragile housing market can affect people's decision to move to new houses because of the lower possibility of missing out on the house. In this way, old suburban neighborhoods gradually lose attraction to the next generation.

Thanks to the booming developments in eastern areas, western parts can be another option for homeowners seeking affordable but well-designed suburban developments. Good residential environments with classical or contemporary designs can satisfy various needs. The case neighborhoods showed the possibility of becoming sustainable neighborhoods in the near future.

4. Aging and changes of essential facilities in housing estates in Hiroshima

In the next case study, we focused on the location of nursing care insurance service projects such as nursing home care and home care facilities, among welfare facilities for older persons in suburban housing estates, and examines the location of nursing care welfare facilities in the suburban housing estates of Hiroshima City [2]. And we will point out the failures of Japanese town planning. As suburban housing estates were originally targeted at child-rearing households, they were rarely established from the very beginning at the time of the development of welfare facilities for older persons. However, owing to the aging of residents, the long-term care insurance service business is now more likely to be located in suburban areas [1].

4.1 Aging in suburban housing estates

Suburban housing estates have rapidly lost their appeal since the 1990s, with potential homeowners showing preference for condominiums in city centers [6]. Suburbanization has continued, and suburbs have started to shrink due to changes in the urban housing market and lifestyles. Hiroshima City has become a typical phenomenon in which the population of suburban housing estates is aging. **Figure** 7 shows that many older persons live in suburban housing estates. In the old suburbs, the majority of residents have become elderly, and the most popular family type has only one or two members [1].

During the early stages of development, the residents' age structure indicated that young nuclear families were dominant. Thirty or forty years later, the household couple aged and their children moved out (**Figure 8**). In these old housing estates, the number of children has decreased significantly and kindergartens and daycare centers for children have been closed. As Yui (2019) pointed out, owing to the outflow of young people and an aging population, shopping streets within suburban housing estate are closing, and the number of vacant houses is increasing. As a result, seniors have started going out to shop outside the area and using home delivery services to purchase food.

4.2 Changes of essential facilities

As Kubo and Yui [1] clarified, some buildings have been changed from supermarkets to nursing homes for the older persons. The number of large nursing homes for the older

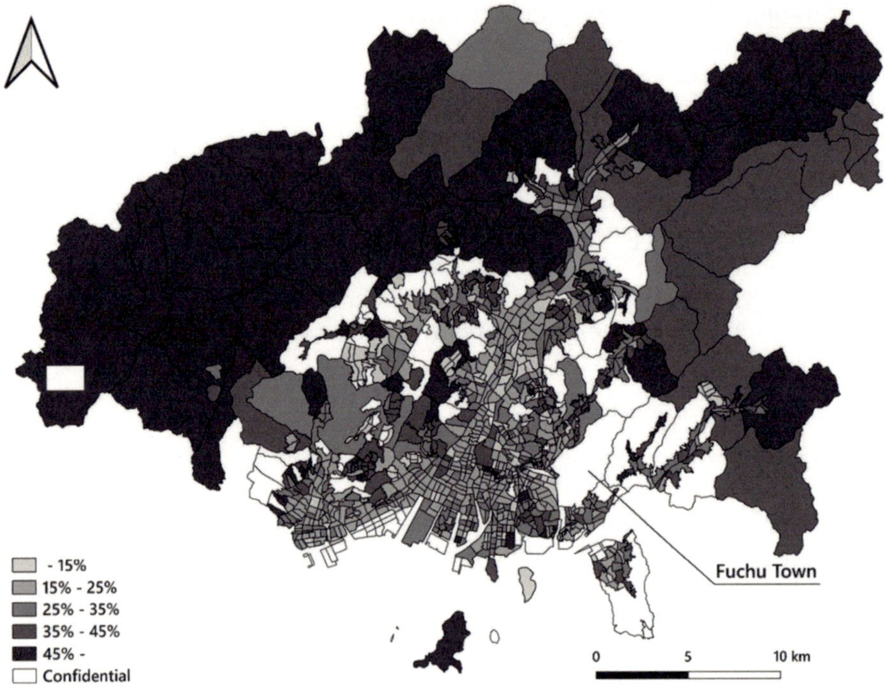

Figure 7.
Rate of older persons in Hiroshima City (2020). Data: National Census.

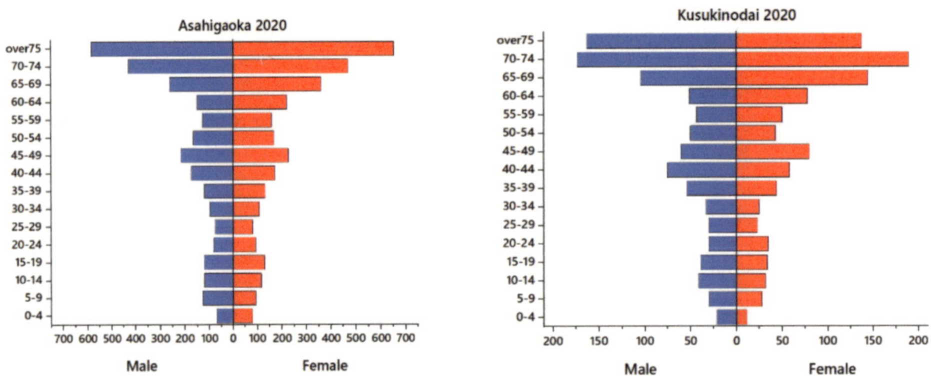

Figure 8.
Age structure in old suburban housing estate in Hiroshima City.

persons is increasing in the suburbs. As the population declines and ages, it has become difficult to maintain shopping malls, and many are closing. Nursing homes for the older persons are located in the suburbs because of the aging population [2]. Aging is proceeding in the inner-city and suburbs. In particular, the number of nursing homes for the older persons is increasing in suburban areas because their children live in distant places and cannot care for their parents. In response to the aging population, welfare institutions for the older persons have increased rapidly in suburban residential areas in recent years. Both home nursing care service offices and nursing home care service offices are widely distributed in residential areas in the suburbs.

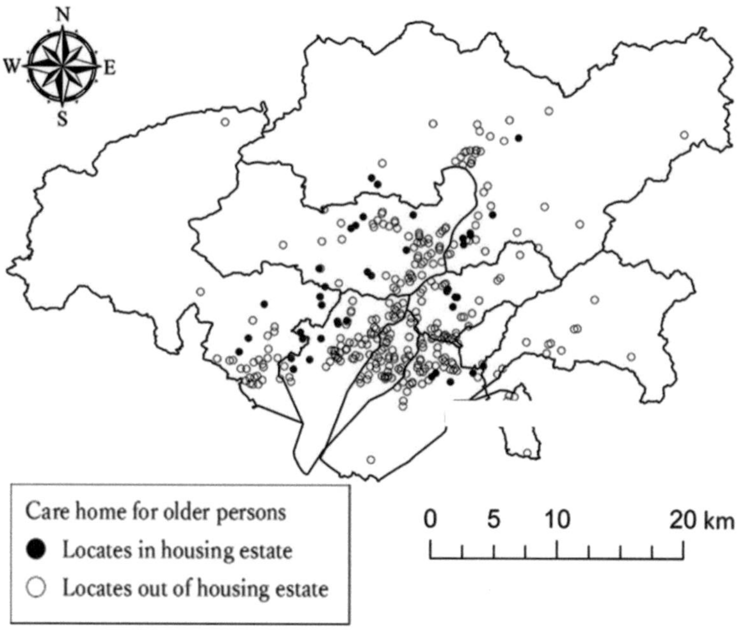

Figure 9.
The locations of nursing home for older persons in Hiroshima (Reprinted from [2], Copyright 2024).

Yui [2] clarified that nursing homes for the older persons are located in the suburbs because of the aging population (**Figure 9**). Aging is proceeding in the inner-city and suburbs. In particular, the number of nursing homes for the older persons is increasing in suburban areas because their children live in distant places and cannot care for their parents. Some buildings have been changed from supermarkets to nursing homes for the older persons. The transformation of shops to welfare facilities is often observed in suburban areas. Otherwise, the number of large nursing homes for the older persons is increasing in the suburbs. Some vacant houses have been renovated into welfare facilities or cafés where the older persons gather each day.

5. Conclusions for a sustainable community

Aging in housing estates was induced by the failure of town planning, which provided a monotonous housing type in the short term. Aging is caused by the aging of the first generation of migrants, and a high rate of aging is caused by the moving out of the second generation. Furthermore, serious aging is caused by residents' homogeneity across generations because of the period of a housing supply shortage [1, 7]. The housing preferences of young households changed from the suburbs to inner cities. Working women prefer urban residences in order to balance their work and home lives. Thus, gentrification is related to female working conditions. Housing projects try to develop conditions to provide care facilities for the older persons and to enable workers to balance the demands of work and family and, furthermore, they are able to make use of women's abilities in the old neighborhoods. These housing projects attempt to develop conditions that provide care facilities for the older persons and enable workers to balance the demands of work and family. It is necessary to set the period of housing development to more than 20 or more long years. Furthermore, it is also important to plan the mixed development, which supply for various households.

Acknowledgements

This study was financially supported by Toyota Foundation (D19-R-0007: PI is Tomoko Kubo) and JSPS (23H00730: PI is Tomoko Kubo). We would like to express my gratitude to those who contributed to this project: Dr. Yuki Iwai and Mr. Haru Usui at University of Tsukuba. We thank Mr. Satoshi Yokogawa and Mr. Taketo Onitsuka for the drawings maps and graphs at Hiroshima University.

Author details

Yoshimichi Yui[1*] and Tomoko Kubo[2]

1 Hiroshima University, Japan

2 University of Tsukuba, Japan

*Address all correspondence to: yyui@hiroshima-u.ac.jp

References

[1] Kubo T, Yui Y, editors. A Rise in Vacant Housing in Post-Growth Japan: Housing Market, Urban Policy, and Revitalizing Aging Cities. Singapore: Springer; 2019

[2] Yui Y. Increasing welfare services and feature of uses in aging suburban housing estates. Annals of Japan Society for Urbanology. 2018;**51**:169-176

[3] Yui Y, Sugitani M, Kubo T. The housing vacancies in suburbs: A case study of Kure City. Urban Geography of Japan. 2014;**9**:69-77. DOI: 10.32245/urbangeography.9.0_69

[4] Ronald R. The Ideology of Home Ownership. New York: Palgrave Macmilan; 2008

[5] Kubo T. Divided Tokyo: Disparities in Living Conditions in the City Center and the Shrinking Suburbs. Singapore: Springer; 2020

[6] Hirayama Y, Izuhara M. Housing in Post-Growth Society: Japan on the Edge of Social Transition. New York: Routledge; 2018

[7] Kubo T, Yui Y, Sakaue H. Aging suburbs and increasing vacant houses in Japan. In: Hino M, Tsutsumi J, editors. Urban Geography of Post-Growth Society. Sendai: Tohoku University Press; 2015. pp. 123-146

Chapter 6

Is Urban Regeneration in Israel a Tragedy or a Common?

Levine Daphna

Abstract

"The tragedy of the commons" illuminates the tension between two opposing forces that are at work in urban society and space during contemporary urban regeneration: mobility and growth on the level of individuals as they maximize their abilities on the one hand, and the decline and deterioration of cities on the other hand. Based on a data-driven analysis of the spatial and economic changes occurring in Bat Yam, a suburban city in Israel's Tel Aviv metropolitan area, as well as on a qualitative study (n = 25) with professionals, this chapter proposes conceiving of the work of professionals leading urban regeneration as a ceaseless effort to balance the commons, but with no sufficient tools or backing. In their eyes, the city had no other economic way out when it commenced the regeneration process more than a decade ago, although it now appears that the apartment owners, the developers, and the political leadership are leading the city into a situation of over-planning and a state of uncertainty. The research contributes mainly to the understanding of the impact of market forces on disadvantaged towns, which are liable to collapse beneath the burden of the need to provide services and overloaded infrastructure as the population density increases.

Keywords: urban regeneration, housing maintenance, displacement, house price inflation, Israel

1. Introduction

According to the theory of "the tragedy of the commons," free access to and unlimited demand for a resource necessarily result in its destruction through over-exploitation. It is a tragedy because all members of the human race, by nature, act to the best of their ability, and in a wholly rational manner, to maximize their profit from the public domain, and in the process are also party to its exploitation until it is too late, and the resource is exhausted. We propose examining contemporary urban regeneration in Israel using the concept of "the tragedy of the commons" to illuminate the tension between two opposing forces that are at work in urban society and urban space in Israel: mobility and growth on the level of individuals, as they maximize their abilities; and the decline and deterioration of cities, particularly the disadvantaged among them, which, as the population density increases, collapse beneath the burden of overloaded infrastructure and the need to provide services.

In our research on urban regeneration, we focused on the city of Bat Yam, Israel, which is currently ranked last among large cities in Israel in terms of quality of life and is simultaneously leading the country in the approval of building requests for urban regeneration [1]. In Israel, urban regeneration is based primarily on privately owned homes; therefore, starting and maintaining the process requires the involvement of apartment owners. In conjunction with developers, they advance the projects involving their apartments to improve their living environment and increase the value of their property [2]. In high-demand areas and on their margins, where developers seek opportunities for profit, the apartment owners choose or are compelled to advance the redevelopment of their apartments along with their neighbors in their buildings. The municipality advances development plans in accordance with its policy and its vision (**Figure 1**).

To provide background, we collected data from a variety of information sources, and we analyzed the spatial and the economic changes occurring in the city of Bat Yam and their demographic implications. In this chapter, we present the findings of qualitative research we conducted with 25 officials involved in the process of urban regeneration in Bat Yam, including planners in the municipal planning department, members of the local and district planning and building committees, assessors, developers, apartment sales reps, and real estate agents. The findings of this research indicate that, in the eyes of the professionals, the city had no alternative economic route more than a decade ago but to commence the regeneration process, although it now appears that the apartment owners, the developers, and the political leadership are leading the city toward a state of over-planning and uncertainty. The professionals engaged in leading urban regeneration are working to balance the commons, but they lack the tools and backing required to do so.

The chapter begins with background on entrepreneurial urban regeneration in Israel followed by a presentation of the research site and methodology. The findings

Figure 1.
A view of Bat Yam. Picture by Roi Boshi, 2020.

section presents the perspective of professionals on the urban regeneration process in Bat Yam. The conclusion discusses the question of whether the process in Israel constitutes a tragedy of the commons.

2. Literature review: private market-driven urban regeneration

The 1980s witnessed the onset of the establishment of a Neoliberal trend in urban regeneration in cooperation with private elements [3] as a way of maximizing the market value of lower value neighborhoods and attracting private investment [4]. The public system assumed that the development of private assets would lead to a rejuvenation of urban spaces [5–7], increased demand for sites in city centers, and improvement of the appearance of the city and the economic activity it contains [8]. Therefore, to an increasing extent, the market has become responsible for the planning, financing, and implementation of regeneration on urban land with high-profit potential, which is typically populated by disadvantaged residents. This trend has been accelerating since the 2008 mortgage crisis in the United States and the intensification of financialization and investment in real estate over the past decade [9].

From the perspective of the private sector, urban regeneration sites in the neglected city centers are considered to be realms of risk and uncertainty, due to the absence of proven demand for new expensive buildings in these areas and to the involvement of many different actors [10]. Regeneration in city centers usually requires government and municipal intervention to stimulate the activity of the market by means of incentives [11]. As a result, public planning policy, which advances urban regeneration, deals primarily with increasing the efficiency of the planning system, accelerating the process of land acquisition, and using planning as a lever for reintroducing run-down sites to the land market [12].

The visual framework regulating the public-private partnership in spot planning is known as the "planning deal" [5, 13, 14]. This is a reference to the agreement between the municipality and the developer, in which the developer takes upon itself planning obligations, public building undertakings, or increased taxation in exchange for building rights [15]. Development based on urban regeneration plans conducted within the framework of planning deals has been the subject of significant criticism. Scholars have been critical of the policy of market-driven urban regeneration as fragmented planning that does not effectively integrate public needs [16]. The phenomenon of large numbers of spot plans and reduced long-term comprehensive planning has been suspected of playing a role in doing injury to public interests [17, 18]. Scholars have argued that this is not a strategy for the development and establishment of infrastructure in urban areas, with the exception of the regeneration of individual housing sites, and has emptied the goals of city planning of all content [19–21].

From a social perspective, gentrification is perceived as an inseparable part of the process, with the goal of bringing investment and members of the middle class back to the central parts of the city, at the expense of the local population [22–25], through "creative destruction" [26]. The research literature identifies a gap between the profit of developers and strong population groups on the one hand, and the injury done to disadvantaged and low-income populations [27] on the other hand; urban redevelopment, it is argued, is an effective tool for accelerating regeneration, but not for achieving equitable development [28], as it involves the dispersion of communities in public housing [29] and detrimentally impacts community cohesion in working-class neighborhoods [30]. Broad discussion has revolved around the original residents' displacement from their homes during processes of urban regeneration [9, 31, 32].

On the other hand, several researchers have argued that the phenomena of gentrification and displacement are not as widespread as people tend to think [32–35]. Freeman [33] suggests that displacement and higher mobility play minor roles as forces of change in gentrifying neighborhoods. According to his research in the urban United States, demographic change in gentrifying areas results from lower rates of intra-neighborhood mobility and the relative affluence of in-movers. Hamnett [35] argues that it may be more appropriate to view the gentrification-induced displacement in London between 1961 and 2001 partly as a replacement resulting from the transition from an industrial to a post-industrial urban economy. Shaw [36] finds that when the gentrification and displacement do occur, they are not necessarily negative, as the displaced move to neighborhoods in which they can find better housing for cheaper rent. Easton et al. [37] argue that in the absence of longitudinal data regarding individual income, it is impossible to say whether the reduction in the number of the most poor in the neighborhood stems from their migration or from the fact that they are benefitting from the socioeconomic improvement of the region [38, 39]. They also maintain that, because the elements associated with Neoliberalism differ from country to country and place to place, they are difficult to use in a generalizing manner in order to understand gentrification-like processes on the local level [40].

In Israel, many apartment owners are party to the process of urban regeneration and therefore profit from their improved living conditions and the increase in the value of their property [41]. Mualam et al. [42] quantified the economic profit of all the actors in projects conducted under National Outline Plan 38 (commonly known by the Hebrew-language acronym "Tama" 38) and showed the many benefits to the apartment owners, in addition to the displacement of renters from Raze-and-Rebuild projects [43]. They also found a direct link between the long-time residents' mode of activity—selling or renting—at Raze-and-Rebuild sites after their regeneration, and real estate values [44]. That is to say, in the Israeli case, the research found different patterns of action than the classic displacement and gentrification [45].

The tension between these two perspectives, the good of the individual, as pursued in urban regeneration in Israel, and the reduction of the overly general view—that is, injury to the general good—led us to ask whether urban regeneration could be examined using the theory of the "tragedy of the commons." It is customary to explain this theory using the analogy of a fishpond in a small village. If each of the village residents fishes to his or her heart's content without taking others into consideration, their extensive fishing would, in practice, be detrimental to the fish's reproduction, and, ultimately, all the residents of the village, including themselves, would go hungry. Even as the urban system is collapsing under the current abundance of individual residential regeneration projects of residents and developers and the need to provide services, this chapter asks whether this is a blind race based purely on the subjective logic that every individual member of the human race by nature acts to the best of their ability to maximize their profits from the public space. In this case, does not the unlimited demand for the resource (urban land) lead to its own exhaustion due to over-exploitation? Will a change in the management of common resources, in a manner that is beneficial to the general public, occur before it is too late? Are Bat Yam and other cities in Israel facing a tragedy of the commons?

3. Methodology

Bat Yam is located on the Mediterranean coast in the first ring of the Tel Aviv Metropolitan Area. It has an area of 816 dunams and a population of 129,000 [46]. For the sake of comparison, Tel Aviv is home to 3.5 times the number of residents and has 6.3 times the area. In other words, Bat Yam is almost twice as densely populated as Tel Aviv. It also has a cumulative deficit and a low average income in comparison with Israel's major cities [47].

The population of Bat Yam leads the poverty indexes *vis-à-vis* the metropolitan and the national average; it is also home to a significant percentage of elderly residents, recipients of state support, and immigrants living in crowded conditions. The population indices for the city (Socioeconomic Cluster 5) reflect a particularly high concentration (20%) of residents over the age of 65, most of which receive state support, and a relatively small percentage of young residents. The rate of academic education is lower than the national average (22.8 versus 28.4%), and the rate of those receiving unemployment benefits and earning minimum wage is higher than the national average. The immigrant population in Bat Yam today accounts for 33% of the total population, which is markedly higher than that of the metropolitan area and double the national average (16%). The percentage of apartment renters is also high, standing at 35.8%, in comparison with a national average of 27.6% (**Table 1**) [48].

Implementation of the Raze-and-Rebuild plans for urban regeneration that are currently being advanced within the planning system will result in the addition of 26,000 housing units to the city's already existing 50,000 units, and a population growth of approximately 50%. In the absence of vacant land, this means that population density will grow accordingly. Some 12,000 old housing units in Bat Yam are currently taking part in urban regeneration, meaning that approximately one-third of the apartments in the city today are involved in a regeneration project of some kind. On average, the number of housing units per lot in Raze-and-Rebuild projects increases fourfold, and under National Outline Plan 38 (the government plan for protecting buildings against earthquakes, including reinforcing the building and adding additional built space to the existing building), it increases by a factor of 1.7. Under National Outline Plan 38/2 (a plan for reinforcing buildings from earthquakes that allows for a building's demolition and re-construction), the number of housing units increases by an average factor of 3.4 (**Figure 2**) [49].

In the course of 2021, we conducted a qualitative study in Bat Yam that included tours in the city and semi-structured interviews (n = 25) with officials and others involved in the planning, approval, marketing, and sale of the apartments in the regeneration projects. We selected the interviewees individually, and we contacted them for this purpose on our own initiative in an effort to understand the public motivations for promoting redevelopment in the city. This included 10 past and present managers and employees from the municipal engineering department and the urban regeneration administration: 3 assessors and economists, 2 members of the district and local planning and building committees, 2 planners from the neighboring municipality of Tel Aviv, 1 municipal welfare worker, and 7 developers, sales reps, and real estate agents. The interviews were recorded, transcribed, and thematically analyzed.

	Criterion	Bat Yam	Holon	Rishon Lezion	Tel Aviv	Nationwide
Demog. Profile	Avg. Household Size	2.6	2.8	3.1	2.1	3.2
	% Age 0–17	20	25	25	19	33
	% Age 65+	19	15	11	14	10
	% 1990 Immigrants + of Total Pop.	33	17	18	13	16
Edu. and Employment	% with Academic Education	19	20	24	37	23
	% with Matriculation Cert.	23	27	27	25	23
	% in Civic Workforce	60	64	68	70	60
	% of Employees Earning up to Min. Wage	32	27	25	27	31
	Avg. Wage of Employees (NIS), 2010	6,938	8,621	9,896	10,837	7,921
	% Recipients of Unemployment and Income Sec. Benefits, 2011	2.4	1.4	1.5	1.6	
Ownership of Sustainable Products	% of Households with at Least One Car	48	66	74	60	62
	% of Households with Two or More Cars	8	20	29	17	19
	% of Households with a Personal Computer	64	73	81	79	71
	% of Households with Internet Service	94	94	96	95	91
Housing	% Living in Owned Apt.	64	71	73	45	66
	% Living in Rented Apt.	33	26	24	48	26
Socio-Econ. Measures	Socio-Economic Cluster	6	6	7	8	
	Dependency Ratio (per 1000)	716	728	641	549	838
	Gini Coefficients	0.37	0.41	0.43	0.48	0.39

Source: Central Bureau of Statistics, Local Authorities in Israel, 2011.

Table 1.
Characteristics of Bat Yam residents in comparison with neighboring cities before implementation of urban regeneration.

4. Findings: Tragedy of the commons?

4.1 A life-saving medication

The study interviewees pointed out the fact that the point of departure of the planning work in Bat Yam, when urban regeneration got underway approximately

Figure 2.
An example of an urban regeneration site, among many others. Picture by the author, 2022.

15 years ago, was that construction and development constituted the life-saving medication for the city. They said that without engaging in construction and demolition, the city would continue to decline and deteriorate. We heard that "in another two decades there will be twenty million people here. Hundreds of thousands of housing units will need to be added to the Tel Aviv metropolitan area in the coming years, and the crowding will continue to increase" (L.) The thinking was that Bat Yam should take full advantage of the construction: "This will allow us to leverage processes of change of which the city is in need. If we do not do so, the infrastructure will remain as it is, the buildings will remain and continue to deteriorate, and the disparity between Bat Yam and other cities will only continue to grow" (N.).

In Bat Yam, the way to engage in regeneration was to re-plan and rebuild: "It is not without its problems, but it is preferable to leaving the status quo" (P.), as "what exists is not good. There are many problems with the status quo, in infrastructure, in people's lives, in the city's ability to be what we would have liked it to be. The way to change it is through the implementation of urban regeneration" (N.). Urban regeneration, the interviewees maintained, is "the only existing alternative" (L.). Therefore, striving for improvement was perceived to be a worthy aspiration, and city planning was the main tool for doing so.

4.2 Striving for the common good

Urban regeneration in Bat Yam has been based on strategic plans that were approved in 2009 and that constituted the main working document. The strategic plan asked the following questions: What is the municipal capacity? How should it be distributed among the neighborhoods? And, what scope should be planned for in order to reach it? One conclusion reached by the team of professionals working on the plan was the fact that, although Bat Yam is one of the most densely populated cities in Israel, it is not densely populated enough: "That is to say, its density was not good enough, and it was crowded with people, but nothing else" (L.). The professionals viewed urban density as a planning ideal with economic and environmental justifications as well. The interviewees used density to describe efficiency in the management of city infrastructure and resources. As a result, the plan called for increasing density in the most deteriorated places in the city that are in need of regeneration more than others, or where there is already enough infrastructure, such as public or regional institutions located close to main transportation arteries, or where development could stimulate the city's economy.

Urban regeneration was a tool for improving existing infrastructure, physical and social alike. From the perspective of the local municipality, it was an opportunity to create more attractive public spaces, to bring more public transportation to the city, to renovate the city center that had atrophied, and "to upgrade built environments that for years had been in a run-down state" (A.). At the same time, some of the interviewees noted that the planning team's Neourbanist approach to compact urbanism fit like a glove: "We told ourselves that we were not simply destroying the neighborhood, and we justified it, from a planning perspective, as an improvement to the quality of city life" (S.).

A main principle of urban regeneration has been to plan as much space for employment and commercial use as possible, in order to offset the additional housing units in terms of local property tax revenue (*arnona*). In Israel, the taxation of employment space is the primary tool possessed by municipal authorities to improve the urban economy, as opposed to residential space, which is deficit. As the city's economic situation was poor to begin with, the planning aspiration was to produce employment space (S.). However, in the conditions that existed when the plan got underway, it was extremely difficult to attract investors to build office buildings in the city, and therefore, large amounts of residential space were also added: "If there is a city that is in good economic condition, this may raise the question: Why spoil it with more housing units? But part of what we wanted to do was to change the balance" (N.). Another argument was that the regeneration of housing would bring employment, "as it is a complete image package, and the more new buildings there were, the greater the chance the office buildings will have to be built" (O.).

4.3 The new social

The interviewees repeated the fact that the municipality's situation had been nothing to speak of from the outset and that the situation of the residents, the owners, and the renters alike was also not good: "Life in buildings that are falling apart, with terrible services and a chronic lack of funding, is not a good life. People sign and are part of the processes because they do not like the status quo" (N.). Urban regeneration made it possible to improve the state of residential buildings that were in poor condition and that residents were unable to maintain and a municipality without resources (L.). In terms of private profit, the apartments taking part in the regeneration projects will enjoy a new or renovated reinforced building with an elevator and an enlarged apartment. In addition, regeneration is also perceived as being beneficial in that it "lifts up the people economically" (S.). That is, the interviewees pointed out that they regard the planning work as making a social contribution by both improving the living conditions of the residents and raising the value of their property: "It is a different conception of social" (S.). Moreover, the interviewees regarded the positive impact of regeneration as also influencing the neighborhood residents whose homes are not undergoing regeneration, "because they create new surroundings" (L.), or, in other words, because "some earn more than others, but no one gets left behind" (N.), and "because it is unavoidable that, ultimately, the buildings that have not yet gotten organized will also undergo regeneration" (S.).

On the other hand, the interviewees were also aware of the extent of the injury sustained during the process and said that, in practice, the distribution of rights in the city is not carried out equitably, but rather depends on the self-organization of the residents vis-à-vis the developers. The buildings that advance in the process receive the construction rights, and those that do not do so suffer from a blockage of light and air: "They will dig them in from all sides, and the building is no longer in its early days, it is already falling apart itself. There is real concern here" (R.). In other words, they told us that "regeneration makes shade, but not on the person who receives the apartment, rather on those who do not" (S.), that economic motivations were stronger than the need to reinforce the buildings, and that the entire population loses more than it gains: "It is an accident of sorts over which market forces took control. There is nothing here beyond the tool for making a profit, and the general population pays for it on all possible levels. The vacuum from which the planners emerged was entered by market forces" (H.).

4.4 Regeneration "on speed"

Over the years since the approval of the strategic plan, most of the planning work in Bat Yam has focused on the question of whether spot planning on the ground is coordinated with the overall framework. Due to the relatively low land values in the city, the market needed to be stimulated. Therefore, in practice, the interviewees said, almost everything that made it onto the table was advanced: "The attempt was to march this city to become a wheel that would start to move on its own and produce attractiveness to create an engine that works" (P.). The assumption was that the majority would be unable to advance toward implementation and that it would therefore be necessary to approve many so that some were implemented. When the city began a planning drive, the question of when to stop approving plans, or "what should be done on the day it is necessary to say no" arose (N.).

Unlike the Raze-and-Rebuild plans that are implemented with high intensity construction that is necessary to bring about a fundamental change, projects conducted under National Outline Plan 38/1 are limited to reinforcing the existing buildings. Although the possibility of reinforcement could have been an effective tool for maintaining the existing texture and effecting a more equitable division of development, the feeling was that urban regeneration in such a city was "on speed" (R.). Almost all reinforcement projects are approved, and "the feeling is very chaotic, like a pressure cooker" (S.). The interviewees said that "Tama at the existing intensiveness is destructive" (A.), that the city was not prepared for large numbers of constructions sites in dense areas, and that requests for permits are not integrated with one another. The outcome is that "there is not enough parking, sufficient space is not left between buildings, the interface with the street is not dealt with as it should be, and the quality of life of the residents suffers" (A.).

The interviewees noted out that the building standards employed in reinforcement projects are lower and the density created is too high. One noted that "these are crazy situations that cannot be accommodated" (R.), and another stated that "ultimately, they receive cages. It's a terrible thing" (A.). In addition, during the process itself, the residents "actually live in dangerous construction sites" (H.). Some interviewees told us that "if people knew ahead of time, they would not start it" (Z.), that "these are poor projects" (M.), and that they cause "not only damage to the commons, but also to the residents' quality of life. Everyone here fell asleep at the switch" (K.). Nonetheless, "it is a situation that cannot be stopped. They are approving plans and they will [continue to] approve plans, because there is no other solution. It is an extremely unsuccessful building solution provided for a population that is not managing to maintain the building" (D.). Ultimately, however, the interviewees stated that "our ability to intervene on the level of property is not great" (M.), and that "it can be explained to them as much as they want, but they are dealing with the developer, and the developer hardly sees them" (P.).

4.5 The private good versus the public good

The interviewees said that when the team of professionals suggests setting rules for the plans, a mixture of uses, height limitations, and conditions for the public good, the residents usually complain to them that they are delaying implementation: "First and foremost, they are looking at their apartment, and they get annoyed at anything delaying them" (M.). The interviewees frequently hear residents say things like: "Why are you even asking us? Do Raze-and-Rebuild. Don't ask the people, because this will only delay the plans. Raze everything" (L.). But when the professional team proposes a more minor regeneration alternative, such as the municipality itself implementing a Raze-and-Rebuild plan that does not increase the area of their apartments or allocate space for private shops, with the aim of reducing the built mass and the height, or the municipality leading the project, as opposed to a private developer, the residents say: "What are you talking about? The developer already promised us an additional 40 square meters per apartment" (L.). That is to say, the interviewees hold that the result is a mixture of the Neoliberal approach of "let me maximize my profit" (R.) on the one hand, and the residents' expectation that the municipality will assume responsibility for the whole thing, safeguarding the surrounding area and the occupants, and, most importantly, not allowing the apartment owners to end up with bad developers.

Some of the interviewees stressed the involvement of politicians in the work of the professional staff as adding to the chaotic situation. "The politicians tell us: you recommend, but we decide. This weakens us vis-à-vis the residents" (B.). We heard the assertion that planning today is motivated by "the developers' pressuring politicians, and the pressure of residents who understand the game" (H.); that "anything goes and there are almost no rules" (Y.); and that "one hand washes the other" (M.). The interviewees repeated the fact that uncertainty is the hardest part of planning behavior in the city for all parties. "The management frequently changes its mind according to different interests and does not provide backing to the professional teams in a uniform manner. This creates a dynamic nature that is not healthy for the city" (P.). Without the approval of clear policy documents, no planning certainty can be established, and this is detrimental to the proper functioning of the processes of change and creates an unrealistic torrent of expectations.

The interviewees also made arguments against the state, saying that there was no economic support for urban regeneration like there was for the establishment of new settlements. Regeneration could be organized, limited in scope, and better suited to the urban capacity and not come at the expense of the rights of residents. They complained of a severe shortage of human resources and jobs, resulting in an ongoing dynamic of putting out fires. Some focused on the fact that there is no correspondence between the required infrastructure and the intensity of the development: "The public puts pressure on the housing, but in the end there will be no infrastructure, and people will say – someone here didn't do their job" (H.). This quote reflects the conviction that, on the metropolitan level, employment and services should be distributed in a manner that reduces travel and balances the local economy. "In the existing system, however, every city plans for itself according to its [own] needs" (H.), without an integrated vision. This was the reason the city began the massive process of urban regeneration from the outset.

5. Discussion and conclusions

As a result of higher housing density and the lack of income-generating space, in addition to a reduction in government funding in recent decades, Bat Yam is in an ongoing state of being unable to function. The stronger population has left the city, and the built environment has continued to decline. In this situation, the city's management regards private market-driven urban regeneration as the only way of improving living conditions, attracting a young population, and effecting a flow of income into the municipal treasury. The municipality works tirelessly to achieve this goal. However, as we have learned from this study, in the event of a chronic shortage of human resources and an inability to shoulder all the expenses of the anticipated investments in infrastructure, services, and public purposes, the success of regeneration depends on restraint.

As a result of the sharp rise in housing costs in Israel over the past few years, real estate developers have identified the profit potential and have forcefully pushed ahead with urban regeneration. Along with them, the apartment owners in Bat Yam's run-down neighborhoods have been provided with an economic springboard. Local politics play a role in this state of affairs, and a feeling of planning uncertainty prevails in the city. Many plans and permits are on the table concurrently, and the array of considerations is not devoid of external pressures. Although regeneration got underway with an organized strategic plan, this study shows that, in practice, Raze-and-Rebuild plans are not integrated, in terms of scope and location, with the permits issued under National Outline Plan 38, which in themselves create low-quality construction situations. In addition, the pace of development of the necessary infrastructure does not constitute a condition for the approval of plans, creating real concern that the city's anticipated population growth will not be provided for properly.

With regard to the question of whether this is "a tragedy of the commons," in which each individual maximizes their profit from the public good until the point that it harms their chances and their quality of life, we offer answers on three levels: the level of individual building, the city level, and the level of the metropolitan area.

On the level of the individual building, residents seek to maximize their profit. Even if they are aware of the lack of balance *vis-à-vis* their neighbors in the surrounding buildings and the pressure on the public infrastructure that will result, each of them continues to seek their profit. Moreover, they even prefer larger projects with a broader impact, as these projects promise greater certainty and a greater return. The municipality is expected to balance the needs of everyone, but when this is done, municipal employees are accused of holding up the advancement of the plans and imposing too many requirements.

On the city level, urban regeneration is perceived not as a war over limited resources, but rather as a product of the resource. We can compare the situation to a fish pond: when the regeneration began approximately 15 years ago, the pond had already been emptied, and the goal of the urban regeneration was to cause water to flow back into it to bring it back to life. The professional officials in the city are certainly aware of their role of balancing the different desires and acting for the public good. However, this is precisely why they are being accused by the public. Therefore, the pressure of the developers, the residents, and the politicians is leading the city into a chaotic state of uncertainty.

On the level of the metropolitan area, in the absence of integrated aims, capacities, and the division of income among the cities, every city must take care of itself, like on the free market. As we have seen, this approach leads the city into an immense whirlpool of pressures with high costs.

This study indicates that the major points requiring attention when promoting urban regeneration are as follows: first, integration and coordination between the planning and the licensing departments to achieve balance among the various plans; second, an increased quantity of human resources in the municipalities; third, strengthening the professional authority of the planning teams; fourth, creating certainty through the approval of policy documents or an overall plan for urban regeneration; fifth, integration between the needs of the cities in the metropolitan area and the equitable distribution of income, building intensity, and density; and sixth, compatibility in the time table between the required infrastructure on the national level and development on the ground.

The phenomenon described here, with local changes, is consistent with processes of urban regeneration in other cities in Israel, albeit at a lower intensity. In practice, it is apparently characteristic of additional professional systems in the country, which lack the backing and funding of the state. Therefore, in an ongoing state of change and uncertainty, as is typical of urban regeneration, the conclusions of this study should be taken into consideration in order to ensure that Israel is not facing a tragedy of the commons.

Author details

Levine Daphna
Jacobs Technion-Cornell Institute, Cornell University, New York, NY, USA

*Address all correspondence to: dlbarch1@gmail.com

References

[1] CBS. Well-Being Indicators in the Big Cities, 2019 [Internet] 2019. Available from: https://www.cbs.gov.il/en/mediarelease/Pages/2020/Well-Being-Indicators-in-the-Big-Cities-2019.aspx [Assessed: August 2, 2021]

[2] Levine D, Aharon-Gutman M. Cities on the edge: How Bat Yam challenges the common social implications of urban regeneration. Journal of Urban Design. 2022. Available from: https://www.tandfonline.com/doi/abs/10.1080/13574809.2022.2154645

[3] McCarthy J. Partnership, Collaborative Planning and Urban Regeneration. [Internet]. Ashgate Publishing, Ltd.; 2007. Available from: https://www.google.com/books?hl=iw&lr=&id=xwFlZ0YVB_YC&oi=fnd&pg=PP11&dq=McCarthy,+J.+(2007).+Partnership,+collaborative+planning+and+urban+regeneration.+Aldershot:+Ashgate.&ots=FkV_jpgehi&sig=sbeZpNx-xaKgoRgbaj5NftUwLJI [Assessed: November 2, 2020]

[4] Verhage R. Renewing urban renewal in France, the UK and the Netherlands: Introduction. Journal of Housing and the Built Environment. 2005;**20**(3):215-227. DOI: 10.1007/s10901-005-9015-4

[5] Rosen G, Walks A. Castles in Toronto's sky: Condo-ism as urban transformation. Journal of Urban Affairs. 2015;**37**(3):289-310. DOI: 10.1111/juaf.12140

[6] Hyra DS. Conceptualizing the new urban renewal: Comparing the past to the present. Urban Affairs Review. 2012;**48**(4):498-527. Available from: http://uar.sagepub.com

[7] Levine-Schnur R, Ferdman A. On the just distribution of land use rights. Canadian Journal of Law and Jurisprudence. 2015;**28**(2):317-342

[8] Singhal S, Berry J, McGreal S. A framework for assessing regeneration, business strategies and urban competitiveness. Local Economy Journal Local Economic Policy Unit. 2009;**24**(2):111-124. DOI: 10.1080/02690940902717147

[9] Aalbers M. The variegated financaialization of housing. International Journal of Urban and Regional Research. 2017:542-554. Available from:. DOI: 10.1111/1468-2427.12522

[10] De Sousa C. Brownfield redevelopment versus greenfield development: A private sector perspective on the costs and risks associated with brownfield redevelopment in the greater Toronto area. Journal of Environmental Planning and Management. 2000;**43**(6):831-853. DOI: 10.1080/09640560020001719

[11] Adair A, Berry J, McGreal S. Financing property's contribution to regeneration. Urban Studies. 2003;**40**(5–6):1065-1080. DOI: 10.1080/0042098032000074326

[12] Healey P. Urban regeneration and the development industry. Regional Studies. 1991;**25**(2):97-110. DOI: 10.1080/00343409112331346327

[13] Levine D, Aharon-Gutman M. The social deal: Urban regeneration as an opportunity for in-place social mobility. Planning Theory. 2022. DOI: 10.1177/14730952221115872 [Assessed: October 29, 2022]

[14] Margalit T, Alfasi N. The undercurrents of entrepreneurial

development: Impressions from a globalizing city. Environment and Planning A Economic Sp. 2016;**48**(10): 1967-1987. DOI: 10.1177/0308518X16651872

[15] Geva Y, Rosen G. The regeneration deal: Developers, homeowners and new competencies in the development process. Geoforum. 2018;**96**(July):10-20. Available from:. DOI: 10.1016/j.geoforum.2018.07.011

[16] Alfasi N. The coding turn in urban planning: Could it remedy the essential drawbacks of planning? Planning Theory. 2018;**17**(3):375-395

[17] Alfasi N. Planning policy? Between long-term planning and zoning amendments in the Israeli planning system. Environment & Planning A. 2006;**38**(3):553-568. Available from: https://journals.sagepub.com/doi/abs/10.1068/a37335?casa_token=d9QEseHm0NwAAAAA:R3LAJzDTsh4xYNSOTuhdYfifwi0ITBaoOfcOTB9bQ0AiEdvQoQi3pxc0BvEH0keaN7E2Yf_wxJsY

[18] Booth P. Controlling Development: Certainty and Discretion in Europe, the USA and Hong Kong. Vol. 9. Psychology Press; 1996. Available from: https://scholar.google.co.il/scholar?hl=iw&as_sdt=0%2C5&q=Booth%2C+P.+%281996%29.+Controlling+development%3A+certainty+and+discretion+in+Europe%2C+the+USA+and+Hong+Kong+%28Vol.+9%29.+Psychology+Press.&btnG= [Assessed: July 7, 2020]

[19] Alexander ER. There is no planning—Only planning practices: Notes for spatial planning theories. Planning Theory. 2016;**15**(1):91-103. DOI: 10.1177/1473095215594617

[20] Purcell M. A new land: Deleuze and Guattari and planning. Planning Theory & Practice. 2013;**14**(1):20-38. Available from: https://www.tandfonline.com/action/journalInformation?journalCode=rptp20

[21] Margalit T. Multi-spot zoning: A chain of public–private development ventures in Tel Aviv. Cities. 2014;**37**: 73-81. Available from:. DOI: 10.1016/j.cities.2013.12.001

[22] Butler T. For gentrification? Environment and Planning A Economic Sp. 2007;**39**(1):162-181. DOI: 10.1068/a38472

[23] Hackworth J. Postrecession gentrification in New York city. Urban Affairs Review. 2002;**37**(6):815-843. DOI: 10.1177/107874037006003

[24] Harris A. From London to Mumbai and back again: Gentrification and public policy in comparative perspective. Urban Studies. 2008;**45**(12):2407-2428. DOI: 10.1177/0042098008097100

[25] Smith N. New globalism, new urbanism: Gentrification as global urban strategy. Antipode. 2002;**34**(3):427-450. DOI: 10.1111/1467-8330.00249

[26] Weber R. Extracting value from the city: Neoliberalism and urban redevelopment. Antipode. 2002;**34**(3): 519-540. DOI: 10.1111/1467-8330.00253

[27] De Oliver M. Gentrification as the appropriation of therapeutic "diversity": A model and case study of the multicultural amenity of contemporary urban renewal. Urban Studies. 2016; **53**(6):1299-1316. DOI: 10.1177/0042098015576314

[28] Chapple K. Planning Sustainable Cities and Regions: Towards More Equitable Development. London: Routledge; 2015

[29] Stevens CA. Public housing redevelopment as social policy. Urban Policy and Research. 1995; **13**(2):81-88. DOI: 10.1080/08111149508551659

[30] Kim H, Marcouiller DW, Choi Y. Urban redevelopment with justice implications: The role of social justice and social capital in residential relocation decisions. Urban Affairs Review. 2018; **55**(1):288-320. DOI: 10.1177/1078087418759605

[31] Ding L, Hwang J, Divringi E. Gentrification and residential mobility in Philadelphia. Regional Science and Urban Economics. 2016;**61**:38-51. Available from: https://www.sciencedirect.com/science/article/pii/S0166046216301223

[32] Freeman L, Braconi F. Gentrification and displacement New York city in the 1990s. Journal of the American Planning Association. 2004; **70**(1):39-52. DOI: 10.1080/01944360408976337

[33] Freeman L. Displacement of succession? Residential mobility in gentrifying neighbourhoods. Urban Affairs Review. 2005;**40**(4):463-491. DOI: 10.1177/1078087404273341

[34] Vigdor JL. Does Gentrification Harm the Poor? Brookings-Wharton Papers on Urban Affairs. Brookings Institution Press. 2002. pp. 133-182. Available from: https://www.jstor.org/stable/25067387?casa_token=KJMx-6nL1lUAAAAA%3AyuclMSRlZ_v_frZhoU1at-4p_qpDogdOYhdVWl-3ST5-exujFe5GRMlesIAm-vniKfjeutmc1KDMIR-T3Pzbnt7UwTBMu4QTcZTw2cwHzaFOei8AqY0&seq=1

[35] Hamnett C. Gentrification and the middle-class remaking of inner London, 1961-2001. Urban Studies. 2003;**40**(12): 2401-2426. DOI: 10.1080/0042098032000136138

[36] Shaw K, Hagemans IW. Gentrification without displacement and the consequent loss of place: The effects of class transition on low-income residents of secure housing in gentrifying areas. International Journal of Urban and Regional Research. 2015; **39**(2):323-341. DOI: 10.1111/1468-2427.12164

[37] Easton S, Lees L, Hubbard P, Tate N. Measuring and mapping displacement: The problem of quantification in the battle against gentrification. Urban Studies. 2019;**57**(2):286-306. DOI: 10.1177/0042098019851953

[38] Atkinson R. Measuring gentrification and displacement in greater London. Urban Studies. 2000; **37**(1):149-165. DOI: 10.1080/0042098002339

[39] Ellen I, O'Regan K. How low income neighborhoods change: Entry, exit, and enhancement. Regional Science and Urban Economics. 2011;**41**:89-97. Available from: https://www.sciencedirect.com/science/article/pii/S0166046211000044?casa_token=4OX6yuqeSakAAAAA:xwpRO91W-ONa4VAizuFrujqrBUZ_NXCsxbN4TKCFk02EpbOz-ePvjeWd0oDBa_8vPp9_5rB8

[40] Kleinhans R, Kearns A. Neighbourhood restructuring and residential relocation: Towards a balanced perspective on relocation processes and outcomes. Housing Studies. 2013;**28**(2):163-176. DOI: 10.1080/02673037.2013.768001

[41] Levine D, Sussman S, Yavo Ayalon S, Aharon-Gutman M.

Rethinking gentrification and displacement: Modeling the demographic impact of urban regeneration. Planning Theory and Practice. 2022:578-597. DOI: 10.1080/14649357.2022.2117399

[42] Mualam N, Salinger E, Goldberg S. Implementing Value Capture in Israel: An Examination of Recent Tools and Policies for Urban Renewal and Earthquake Preparedness [Internet]. 2020. Available from: https://www.lincolninst.edu/publications/working-papers/implementing-value-capture-israel [Assessed: November 2, 2020]

[43] Kainer PN. Housing regeneration strategies: Evaluation from a social equity point of view (Hebrew). Technion. 2017

[44] Geva Y, Rosen G. Socio-spatial outcomes of raze and rebuild ('pinui-binui') projects (Hebrew). Plan. 2019;**16**: 201-224

[45] Levine D, Aharon-Gutman M. There's no place like real estate: The “self-gentrification” of homeowners in disadvantaged neighborhoods facing urban regeneration. Journal of Housing and the Built Environment. 2022;**37**(2): 1-20. DOI: 10.1007/s10901-022-09970-0

[46] CBS. Statistical Abstract of Israel, 2020—No 71. [Internet] 2020. Available from: https://www.cbs.gov.il/he/publications/Pages/2020/שנתון-סטטיסטי-לישראל-2020-מספר-71.aspx [Assessed: February 15, 2021]

[47] Eizenberg E, Cohen N. Reconstructing urban image through cultural flagship events: The case of bat-yam. Cities. 2015;**42**(PA):54-62. Available from:. DOI: 10.1016/j.cities.2014.09.003

[48] CBS. The Local Municipalities in Israel [Internet]. 2018. Available from: https://www.cbs.gov.il/he/publications/Pages/2020/הרשויות-המקומיות-בישראל-2018.aspx#losExcelos [Assessed: July 28, 2021]

[49] Levine D, Sussman S, Aharon-Gutman M. Spatial-temporal patterns of self-organization: A dynamic 4D model for redeveloping the post-zoning city. Environment and Planning B: Urban Analytics and City Science. 2022;**49**(3): 1005-1023. DOI: 10.1177/23998083211041369

Chapter 7

Exploring the Geo-Tourism Potential and Its Accessibility in Danube Region Serbia: A Geo-Statistical Approach

Ana Vulevic, Stabak Roy, Rui Alexandre Castanho, Mara Franco and Gualter Couto

Abstract

Geo-tourism, an emerging field that focuses on the natural and cultural heritage of a region, offers a unique opportunity to promote sustainable tourism and foster local economic development. This study aims to assess the geo-tourism potential Danube region in Serbia, a natural diverse and culturally rich region of Serbia, Western Balkan, and Southeastern Europe, using a comprehensive methodology that incorporates geo-statistical and machine learning tools. A dataset comprising various geographical, and cultural factors was collected from reliable sources, including, protected areas, tourism statistics, cultural heritage inventories and satellite imagery. Geo-statistical analyses were performed to identify spatial patterns and relationships among the collected variables. Techniques such as spatial autocorrelation, hotspot analysis, and interpolation methods were employed to reveal concentrations of geo-tourism resources, hotspots, clusters, and areas in need of conservation. The results of this study provided valuable insights into the geo-tourism potential of the Danube region. The spatial analysis revealed several hotspots. Machine learning models accurately predicted tourism demand based on variables such as accessibility, cultural heritage, and natural landscapes. These findings can guide policymakers that, using the power of geo-statistical and machine learning tools, the Danube region in Serbia can unlock its full geo-tourism potential.

Keywords: geo-tourism, Danube, geo-statistical analysis, accessibility, machine learning, sustainable tourism, cultural heritage, natural protection area

1. Introduction

Geo-tourism has emerged as a novel idea within the realm of tourist management, garnering considerable scholarly and industry interest in recent times [1]. This perspective provides a novel approach to the concept of tourism, highlighting the significance of safeguarding local identities, maintaining environmental integrity, and fostering sustainable economic growth [2]. The objective of this method is to achieve a harmonic equilibrium between the promotion of tourism and the preservation of the unique attributes that contribute to the cultural richness and natural diversity of a particular place [2]. The region of the Danube in Serbia, situated in the Western Balkans and Southeastern Europe, has significant prospects for the development of geo-tourism. The region in question, which has historical significance and showcases cultural diversity, exhibits a distinctive amalgamation of historical sites, scenic landscapes, and lively local traditions. By adopting an appropriate methodology and engaging in meticulous strategic deliberation, the Danube area has the potential to use its resources effectively in order to cultivate sustainable tourism methodologies that safeguard the region's innate natural and cultural legacy [3]. In order to assess the potential for geo-tourism in the Danube area of Serbia, a thorough approach will be utilised. The proposed technique would entail the use of geo-statistical analysis and advanced machine learning methods to evaluate the strengths, weaknesses, possibilities, and dangers pertaining to tourist growth in the region. Geo-statistical investigations can offer significant insights into the geological and geographical aspects of the region, facilitating the identification of distinctive features and attractions that can be effectively utilised for tourist purposes [4, 5]. Through the examination of several data sets encompassing landforms, hydrology, climate, and biodiversity, scholars are able to delineate the natural resources present inside a certain place and ascertain their capacity to allure tourists seeking nature-centric encounters [6]. In addition, a comprehensive analysis will be conducted on the cultural heritage of the region in order to identify significant historical sites, traditional artisanal practises, festive events, and other cultural manifestations that may be effectively marketed as integral components of the geo-tourism encounter [7]. The present investigation will encompass an examination of the historical background, natural, folklore, and indigenous customs of the region in order to compile an all-encompassing catalogue of region cultural resources that may be effectively incorporated into the plan for tourist development [8]. Furthermore, alongside the geostatistical analysis, state-of-the-art machine learning techniques will be utilised to evaluate extensive datasets and detect underlying patterns and trends. These techniques facilitate the identification of tourist preferences, behaviour, and travel patterns, enabling tourism planners to customise their products in order to cater to specific requests [9]. By comprehending the requirements and aspirations of prospective tourists, the Danube area has the ability to generate distinctive and lasting experiences that are in accordance with the concepts of geotourism. The primary objective of this research is to provide a comprehensive framework for the advancement of sustainable tourism in the Danube area, with a focus on optimising economic advantages while concurrently mitigating adverse effects on the environment and local populations [10]. The proposed roadmap would encompass suggestions for the advancement of infrastructure, implementation of visitor management tactics, enhancement of marketing and promotional endeavours, and the implementation of capacity-building activities [11]. The efficacy of geo-tourism in the Danube area is contingent upon the active engagement and cooperation of many stakeholders, encompassing local communities, governmental bodies, tourist institutions, and commercial sector organisations. The active involvement of stakeholders and their empowerment in the decision-making process is of paramount importance for the sustainable success of geo-tourism programmes [12]. The area of the Danube in Serbia has significant potential for the development of geo-tourism. Through the implementation of a comprehensive approach encompassing geostatistical studies and state-of-the-art machine learning techniques, the area can effectively discern its inherent natural and cultural resources and use them to cultivate sustainable tourist experiences. By engaging all pertinent stakeholders and upholding the tenets of geo-tourism, the Danube area has the potential to attain a peaceful equilibrium between the promotion of tourism and the conservation of its distinctive history.

2. Methodology

2.1 Statistical elements

The basis of this study relies on a well selected dataset that encompasses a diverse range of geographical and cultural variables. The aforementioned elements were derived from reputable sources records of protected areas, tourism data, encompassing cultural heritage inventories and satellite images [13]. The choice of variables was determined based on their direct pertinence to the field of geo-tourism, specifically emphasizing qualities such as protected areas, historical sites, and tourism infrastructure. Spatial patterns and interrelationships among the gathered data were investigated through the implementation of geo-statistical techniques. By utilizing methodologies such as spatial autocorrelation, hotspot analysis, and interpolation techniques like kriging, the study revealed clusters of geo-tourism assets and identified regions with significant potential for further development. Through the identification of hotspots, clusters, and areas for protection or improvement, this portion of the study has shown a comprehensive comprehension of the region's potential.

Spatial autocorrelation, often denoted as I or I(d) quantifies the degree of similarity or dissimilarity between values of a variable at different spatial locations [14]. It indicates whether similar values tend to cluster together or if dissimilar values are spatially dispersed. In mathematical notation, the spatial autocorrelation index *I* is often computed using Moran's *I* statistic, which is defined as [14]:

$$I = \frac{n}{\sum_{i=1}^{n}\sum_{j=1}^{n} w_{ij}} \cdot \frac{\sum_{i=1}^{n}\sum_{j=1}^{n} w_{ij}.(x_i - \overline{x})(x_j - \overline{x})}{\sum_{i-1}^{n}(x_i - \overline{x})} \tag{1}$$

Where,

n is the number of spatial units (locations);

x_i and x_j are the values of the variable of interest at locations i and j respectively;

$\overline{x}$ is the mean of all x values;

w_{ij} represents the spatial weight between locations i and j.

Moran's I ranges from −1 (perfect dispersion) to 1 (perfect clustering), with values near 0 indicating spatial randomness. The distance-based spatial autocorrelation index, often denoted as \(I(d)\), incorporates a distance threshold I(d) to measure autocorrelation only among locations within a certain distance from each other. It can be expressed as [14]:

$$I = \frac{n(d)}{\sum_{i=1}^{n}\sum_{j=1}^{n} w_{ij}(d)} \cdot \frac{\sum_{i=1}^{n}\sum_{j=1}^{n} w_{ij}(d).(x_i - \overline{x})(x_j - \overline{x})}{\sum_{i=1}^{n}(x_i - \overline{x})} \tag{2}$$

Where,

n(d) is the number of locations within distance d of each location.

w_{ij} (d) are distance based spatial weights that depend on the distance d between locations i and j

Both I and I(d) are used to assess the spatial autocorrelation of a variable and are essential tools in spatial analysis to understand the spatial distribution and patterns of tourism in Danube region.

Calculate Getis-Ord Gi* statistics for each tourist spots. Getis-Ord Gi* has been define as follows:

$$G_i^* = \frac{S_i - M_i.S_i}{\sqrt{\frac{V_i.\text{number of neighbors}) - S_i^2}{\text{number of neighbors}}}} \tag{3}$$

Where:

- The local sum S_i of the variable in a neighbourhood around location;
- The local mean M_i of the variable in a neighbourhood around location *i*;
- The local variance V_i of the variable in a neighbourhood around location.
- The calculated G_i^* values are compared to a standard normal distribution to assess statistical significance. Positive values indicate high-value clusters (hotspots), negative values indicate low-value clusters (cold spots), and values near zero indicate random distribution [15]. Moreover, the research utilized machine learning methods to effectively represent the complex relationships among the discovered factors and the demand for tourism. The study employed several prediction models, such as regression analysis, decision trees, and random forests, to analyses historical tourism data. The objective was to estimate future tourist arrivals and identify the main factors that influence visitor preferences.

2.2 Study area

The Danube region of Serbia is an exceptionally diverse geographical corridor encompassing a unique blend of natural landscapes, cultural heritage, and historical significance. As the second-longest river in Europe, the Danube traverses 588 km through Serbia, shaping the landscape and culture of the regions it touches (**Figure 1**). The Danube Basin in Serbia forms part of the larger Pannonian Plain in the northern part of the country. Flowing from the west at Bezdan to its confluence with the Timok River near the Bulgarian border in the east, the Danube skirts the edges of the Fruška Gora mountains, creating a natural boundary between the Balkan and Carpathian Mountain ranges [16]. The region incorporates a range of ecological zones, from wetlands and forests to rocky outcrops and sandy areas. Historically, the Danube has been a conduit for trade, migration, and cultural exchange, and this is evident in the archaeological and architectural remnants scattered throughout its Serbian stretch. Sites such as the Lepenski Vir, which date back to 7000 BC, and the Roman-era fortresses like the one at Golubac testify to the region's historical depth. Several towns and cities along the Danube in Serbia, such as Novi Sad, Smederevo, and Kladovo, have rich cultural heritages [17]. These urban centers are repositories of Serbian art,

Figure 1.
Location map of the study area.

music, literature, and culinary traditions, making them focal points for cultural tourism. The Danube's riparian zones, especially the Djerdap National Park, are of considerable ecological significance. Home to diverse flora and fauna, these areas offer potential for eco-tourism and bird-watching activities [18].

The primary aim of this research is to explore the geo-tourism potential of the Danube region in Serbia, assessing the accessibility of these sites, and predicting future tourism trends using geo-statistical and machine learning methodologies. By focusing on the interplay between natural beauty, historical richness, and modern accessibility, this study aims to provide a comprehensive understanding of the region's potential as a geo-tourism hub.

2.3 Factors influencing visitor preferences in the Danube region of Serbia

Factors influencing visitor preferences in the Danube region of Serbia include:

1. Natural Landscapes: The diverse wildlife and scenic beauty found in the protected lands of the region serve as an additional draw for visitors.

2. Cultural Heritage: Enriching the region's appeal is its abundant cultural heritage, featuring historical landmarks and culturally significant sites.

3. Accessibility: Visitor preferences may be influenced by the ease of reaching and navigating the region, encompassing factors such as well-connected roads, efficient public transportation, and the availability of accommodations.

4. Tourism Infrastructure: The presence of tourist-centric facilities like hotels, restaurants, information centers, and guiding services can significantly impact tourist arrivals.

5. Geotourism Resources: Unique elements such as geoparks, geological monuments, distinctive terrains, biodiversity, and environmental education programs have the potential to attract geo-tourists seeking specific experiences.

6. Sustainability: The extent to which the region adopts sustainable tourism practices is also a noteworthy factor for visitors, reflecting a commitment to minimizing environmental impact and preserving cultural heritage. In **Figure 1** location map shown of the study area

2.4 Accessibility

Accessibility to the nearest regional centers Belgrade and Novi Sad from 27 centers by road is given in **Table 1**.

NUTS3	Urban centers of NUTS3	1	2
		Beograd	Novi Sad
City of Belgrade	Beograd	-	57
South Backa	Novi Sad	57	-
Danube region	Smederevo	49	100
West Backa	Sombor	124	77
Srem	Sremska Mitrovica	48	61
Central Banat	Zrenjanin	70	35
South Banat	Vrsac	79	109
Branicevski	Pozarevac	54	105
Borski	Bor	149	200

Table 1.
Accessibility from NUTS3 centers to the nearest regional centers by road (travel time in minutes).

2.4.1 Accessibility to the nearest regional centers by rail

The analysis of transportation accessibility of urban centers by rail was carried out on the basis of data on the existing organization of rail passenger traffic on the territory of the Republic of Serbia (**Table 2**).

2.4.2 Urban connectivity

The indicators shows how many cities with more than 50,000 inhabitants are accessible within 60 min of travel by car or train (**Table 3**)

NUTS3	Urban centers of NUTS3	Novi Sad	Beograd
City of Belgrade	Beograd	120*	X
South Backa	Novi Sad	X	120*
Danube region	Smederevo	281	161
West Backa	Sombor	143*	270
Srem	Sremska Mitrovica	92*	219
Central Banat	Zrenjanin	161*	153*
South Banat	Vrsac	232	105*
Branicevski	Pozarevac	349	222
Borski	Bor	697	570

**Direct line.*

Table 2.
Accessibility from the NUTS3 centers to the nearest regional centers by rail (travel time in minutes).

Unit code	Name of NUTS3		Access to city functions by rail	Access to city functions by car
			2011	2011
RS002	West Backa	NUTS3	0,99	0,59
RS004	South Banat	NUTS3	1,08	0,61
RS005	West Banat	NUTS3	1,34	0,12
RS006	South Backa	NUTS3	1,41	0,63
RS007	Sremski	NUTS3	2,58	0,76
RS010	Danube	NUTS3	3,10	0,95
RS011	Branicevski	NUTS3	0,25	0,10
RS014	Borski	NUTS3	0,38	0,18
RS025	Grad Beograd	NUTS3	2,41	1,24

Model output. Sources: RRG Accessibility Model (2012), Road Network: RRG GIS Database (2012), Cities > 50,000. Residents: RRG GIS Database (2012).
© ESPON Source of information: ESPON EGTC the interpretation of ESPON material does not necessarily reflect the opinion of the Supervisory Board of ESPON 2020. ESPON project TRACC [19].

Table 3.
Urban connectivity/accessibility of city functions by car and rail.

- Accessibility of city functions by car/Number of cities with more than 50,000 inhabitants that can be reached within 60 min by car; and
- Availability of urban functions by rail/Number of cities with more than 50,000 inhabitants that can be reached within 60 min by rail.

3. Result and discussion

This study aimed to assess the geo-tourism potential of the Danube region in Serbia, a natural diverse and culturally rich area within the Western Balkans and Southeastern Europe. By employing a comprehensive methodology that integrates geo-statistical and machine learning techniques, this research provided valuable insights into the region's capacity to promote sustainable tourism and foster local economic development.

The initial phase of the study involved the collection of a diverse dataset from reliable sources, encompassing geographical and cultural factors relevant to geo-tourism (**Figure 2**). These factors included protected areas, historical landmarks, tourism statistics, and satellite imagery. By utilizing geo-statistical analyses, the study identified spatial patterns and interrelationships among these variables. Techniques like spatial autocorrelation, hotspot analysis, and kriging interpolation were pivotal in unveiling concentrations of geo-tourism resources and areas with elevated potential for development.

The results of the spatial analysis revealed several key hotspots within the Danube region. These hotspots signify geographic areas characterized by high concentrations of valuable geo-tourism resources, including natural landscapes, cultural sites, and

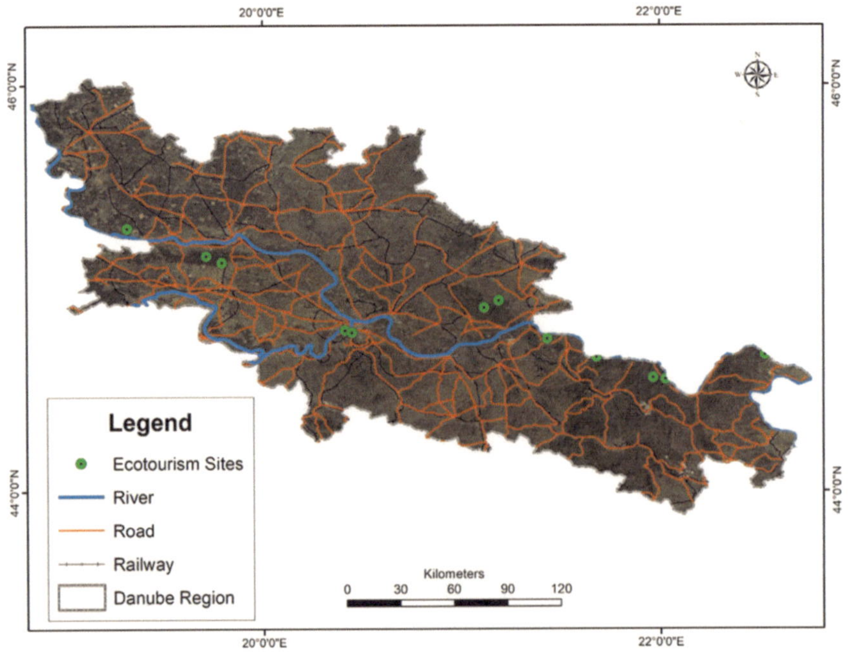

Figure 2.
Major Geo-tourism sites of Danube region of Serbia.

historical monuments. Such insights enable policymakers and stakeholders to pinpoint zones of high significance for tourist attraction, thus guiding the strategic allocation of resources for further development (**Table 4**).

Based on the analysis conducted, it is clear that the geotourism potential of the selected sites in Serbia varies depending on their geological significance, cultural context, accessibility, and recreational offerings. Among the sites assessed, Derdap National Park, Deliblato Sands, and The Iron Gates emerge as particularly promising destinations for geotourism [20]. Derdap National Park, situated along the Iron Gates gorge, showcases unique rock formations, diverse biodiversity, and rich geological features, positioning it as a robust geotourism attraction. Deliblato Sands, with its vast aeolian sand dunes and diverse ecosystems, appeals to nature enthusiasts and researchers due to its geological, botanical, and zoological attributes. Similarly, The Iron Gates gorge, formed by the Danube River cutting through the Carpathian Mountains, offers towering cliffs, intricate stratigraphy, and significant historical importance, making it highly aligned with the principles of geotourism. These three sites exemplify a harmonious fusion of geological features, cultural significance, and ecological attributes, generating a comprehensive and enriching geotourism experience for visitors (**Figure 3**).

In the assessment of geotourism sites within the Danube region of Serbia, a significant pattern emerges as observed in **Figure 4**. A notable 41.66% of the identified geotourism sites are positioned within a 50 km radius of the mean center of tourism activity. This spatial distribution underscores a strategic clustering of these sites in close proximity to the heart of tourism engagement. The spatial concentration of maximum geotourism sites within this specific range signifies a deliberate choice that aligns with the principles of convenience and accessibility. This geographical

Sl. No.	Name	Latitude	Longitude	Tourists (annually)	Distance from river (km)	Distance from road (km)	Distance from railway station (km)
1	Ada Ciganlija	44.8167° N	20.4167° E	3,00,000	0.5	1	1
2	Deliblato Sands	44.9167° N	21.5000° E	1,00,000	10	10	5
3	Đerdap National Park	44.7500° N	21.5000° E	2,00,000	2	5	5
4	Fruškagora	45.1667° N	19.8333° E	5,00,000	5	10	5
5	Golubac Fortress	44.7667° N	21.5500° E	1,00,000	1	2	2
6	Karađorđevo	45.0000° N	21.0000° E	10,000	5	5	5
7	Lepenski Vir	44.8500° N	22.3000° E	1,00,000	3	1	1
8	Srebrnojezero	44.9167° N	20.9167° E	5,000	3	2	2
9	The Iron Gates	44.8833° N	22.2000° E	5,00,000	5	10	10
10	Topčider	44.8167° N	20.4167° E	1,00,000	0.5	1	1
11	Vrdnička Banja	43.4167° N	20.9167° E	5,00,000	5	10	5
12	Zagajička Bara	44.5500° N	21.7500° E	20,000	4	10	10
13	GornjePodunavlje	45°4621″N	18°5537″E	2,50,000	2	1	5

Source: Prepared by the author, 2023.

Table 4.
Major geotourism sites of Danube region, Serbia.

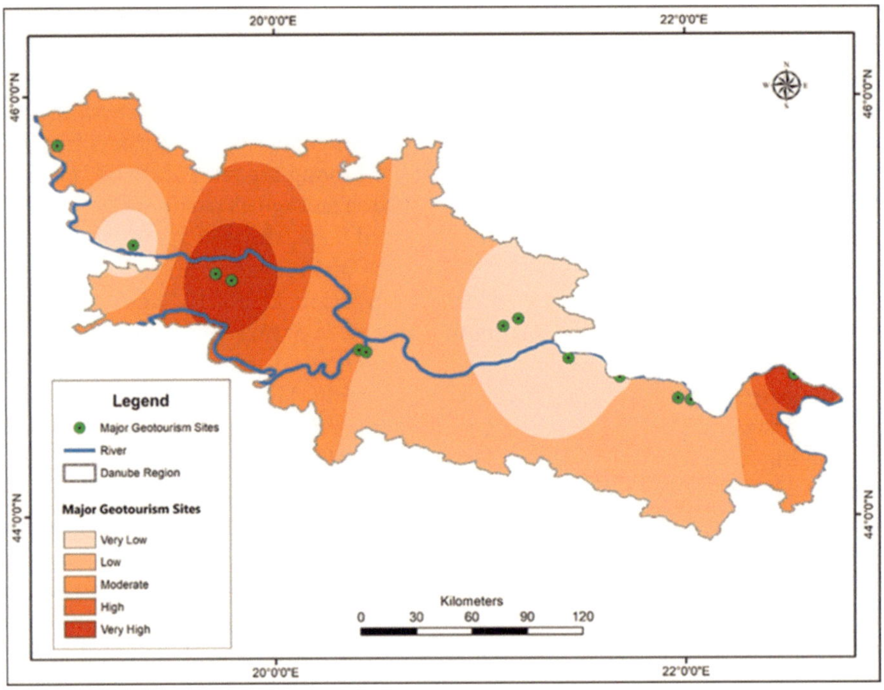

Figure 3.
Geo-tourism potential zone of Danube region of Serbia.

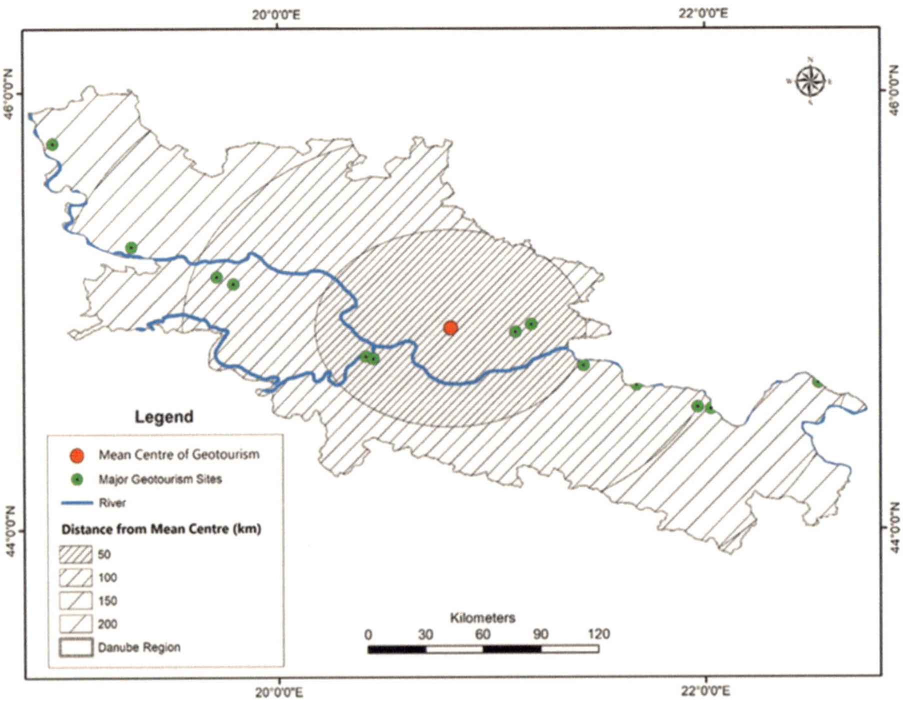

Figure 4.
Mean centre of Geo-tourism potential zone of Danube region of Serbia.

arrangement capitalizes on the advantageous synergy between prominent geotourism destinations and the central tourism hub [21]. The outcome is a seamless fusion of natural and cultural attractions with the well-established tourism infrastructure surrounding the region. One of the driving factors amplifying the geotourism potential in this area is the pronounced influx of potential tourists. The region's allure is magnified by its propinquity to the bustling urban center of Belgrade. The close proximity to this vibrant city plays a pivotal role in attracting a substantial number of prospective visitors. This dynamic relationship between urban dynamism and serene geotourism sites creates a compelling dichotomy, allowing travelers to experience a diverse spectrum of activities within a relatively short radius.

Furthermore, the intrinsic connection between geotourism potential and tourism inflow in this region is buttressed by the remarkable availability of accommodation options. The infrastructure catering to the accommodation needs of tourists is notably abundant in the vicinity of Belgrade. This strategic allocation of lodging facilities ensures that travelers can seamlessly transition between the urban amenities of the city and the immersive geotourism experiences offered by the region's natural and cultural treasures. Notably, accessibility, a cornerstone of sustainable tourism development, has been meticulously enhanced within the Belgrade city region. The well-developed transportation networks and connectivity channels enable swift and effortless movement between the urban hub and the surrounding geotourism sites. This accessibility factor not only enriches the tourist experience but also facilitates the dissemination of economic benefits across the region.

The spatial distribution of maximum geotourism sites within the Danube region of Serbia, predominantly concentrated within a 50 km radius of the mean center of tourism activity, speaks to a strategic alignment of natural and cultural attractions with the well-established tourism core. The symbiotic relationship between urban vibrancy, accessibility, and the allure of geotourism sites showcases the region's ability to offer a multifaceted experience to tourists. This analysis underscores the importance of synergizing infrastructure development, accessibility, and geotourism potential to create a harmonious balance between modern conveniences and immersive natural and cultural exploration.

Building upon the spatial analysis, machine learning algorithms were employed to unravel the intricate interactions between the identified variables and tourism demand. By harnessing historical tourism data, predictive models were constructed, encompassing regression analysis, decision trees, and random forests. These models not only forecasted future tourist arrivals but also illuminated the pivotal factors influencing visitor preferences. The results of the machine learning analysis demonstrated that variables such as accessibility, cultural heritage, and natural landscapes exert significant influence on tourism demand within the Danube region [22]. This understanding enables stakeholders to prioritize initiatives aimed at enhancing accessibility, preserving cultural heritage, and optimizing the natural protection areas to align with visitor preferences.

4. Conclusion

This chapter delves into the emerging field of geo-tourism and its potential to promote sustainable tourism and foster local economic development. By focusing on the Danube region in Serbia, a culturally diverse and rich area [23–26] within the Western Balkans and Southeastern Europe, this study employed a comprehensive methodology that seamlessly combined geo-statistical and machine learning tools. Through the meticulous collection and analysis of various geographical and cultural factors, the study's findings shed light on the region's geo-tourism potential, offering valuable insights for informed decision-making. The integration of geo-statistical techniques allowed for the identification of spatial patterns, clusters, and concentrations of geo-tourism resources. The application of machine learning algorithms further enhanced the analysis by modeling the intricate relationships between variables and tourism demand. Predictive models effectively forecasted tourist arrivals by considering crucial factors such as accessibility, cultural heritage, and natural landscapes. The outcomes of this research hold significant implications for various stakeholders, including policymakers, tourism industry players, and local communities. By leveraging these insights, the Danube region in Serbia can tap into its geo-tourism potential while safeguarding its unique natural and cultural heritage. Overall, this research not only advances our understanding of the geo-tourism potential in the Danube region of Serbia but also showcases the power of interdisciplinary methodologies, specifically the integration of geo-statistical and machine learning tools. As destinations seek to balance economic growth with environmental and cultural preservation, studies like this contribute to the ongoing discourse on sustainable tourism practices, providing actionable insights that can shape the future of tourism development in similar regions worldwide.

The outcomes of this study possess multifaceted implications for the sustainable development of geo-tourism within the Danube region. The spatial analysis, unveiling hotspots and areas of potential, empowers policymakers to concentrate their efforts on infrastructural development, conservation, and enhancement of the identified areas. Simultaneously, the machine learning models provide an avenue to foresee future tourism demand, thereby facilitating more informed decision-making processes.

The integration of geo-statistical and machine learning methodologies in this study underscores their synergistic potential for crafting a comprehensive understanding of the geo-tourism landscape. These tools can be further harnessed by local communities, tourism stakeholders, and policymakers to guide strategic planning, investment, and conservation initiatives. By capitalizing on the identified factors that drive geo-tourism attractiveness, the Danube region can unlock its full potential, nurturing sustainable tourism practices while safeguarding its invaluable natural and cultural heritage. The harmonious fusion of geo-statistical analyses and machine learning models provides a holistic perspective on the geo-tourism potential within the Danube region of Serbia. The insights gleaned from this study pave the way for informed decision-making processes, thereby promoting sustainable tourism growth while preserving the region's rich natural assets.

Acknowledgements

Funded by national funds through FCT – Portuguese Sc.

Conflict of interest

The authors declare no conflict of interest.

Author details

Ana Vulevic[1,2], Stabak Roy[3], Rui Alexandre Castanho[2,4]*, Mara Franco[2,5] and Gualter Couto[2,6]

1 Institute of Transportation CIP, Belgrade, Central Serbia, Serbia

2 CITUR–Madeira–Centre for Tourism Research, Development and Innovation, Funchal, Madeira, Portugal

3 Institute of Socio-Economic Geography and Spatial Management, University of Gdansk, Gdansk, Poland

4 University of Johannesburg, South Africa

5 Madeira University, Portugal

6 Azore University, Portugal

*Address all correspondence to: acastanho@wsb.edu.pl

References

[1] Abou Arrage J, Chamra C. Geo-landscape and geo-heritage assessment to promote geo-tourism and geo-conservation of Ehden region in North Lebanon. International Journal of Geoheritage and Parks. 2022;**10**(4):635-654. DOI: 10.1016/j.ijgeop.2022.10.002

[2] Hadian MSD, Suganda BR, Khadijah ULS, Yuliawati AK. Evaluation of potential geo-tourism based on geo-diversity towards sustainable tourism at citarum river. Estudios de Economia Aplicada. 2021;**39**(12). DOI: 10.25115/eea.v39i12.6265

[3] Antić A, Tomić N, Đorđević T, Marković SB. Promoting Palaeontological Heritage of Mammoths in Serbia through a Cross-Country Thematic Route. Geoheritage. 2021; **13**(1). DOI: 10.1007/s12371-021-00530-6

[4] Gretzel U. Intelligent systems in tourism: A social science perspective. Annals of Tourism Research. 2011;**38**(3): 757-779

[5] Gretzel U, Sigala M, Xiang Z, Koo C. Smart tourism: foundations and developments. Electron Markets. 2015; **25**:179-188. DOI: 10.1007/s12525-015-0196-8

[6] Dragin A, Jovicic D, Lukic T. Cruising along the river Danube: Contemporary tourism trend in Serbia. Geographica Pannonica. 2010;**14**(3). DOI: 10.5937/geopan1003098d

[7] Mikhailenko AV, Ruban DA. Geo-heritage specific visibility as an important parameter in Geo-Tourism resource evaluation. Geosciences (Switzerland). 2019;**9**(4):146. DOI: 10.3390/geosciences9040146

[8] Huang CM, Chen SC. Smart tourism: Exploring historical, cultural, and delicacy scenic spots using visual-based image search technology. Applied Mechanics and Materials. 2015;**764**: 1265-1269

[9] Trišić I, Nechita F, Milojković D, Štetić S. Sustainable tourism in protected areas—Application of the prism of sustainability model. Sustainability (Switzerland). 2023;**15**(6):5148. DOI: 10.3390/su15065148

[10] Bærenholdt JO, Fuglsang L, Sundbo J. A coalition for 'small tourism' in a marginal place: Configuring a geo-social position. Journal of Rural Studies. 2021;**87**:169-180. DOI: 10.1016/j.jrurstud.2021.09.010

[11] Pókó N. River Cruise Tourism Along the Middle Section of the Danube, in Hungary, in Croatia and in Serbia. Acta Economica et Turistica. 2022;**8**(1):25-47. DOI: 10.46672/aet.8.1.2

[12] Milenković J. Evaluation of Geo-sites in the Podrinje-Valjevo Mountains with Respect to Geo-tourism Development. Geoheritage. 2021;**13**(2). DOI: 10.1007/s12371-021-00567-7

[13] Mitra S, Roy S, De SK. Tourism Industry of Namchi, South Sikkim: An overview. Geographical Review of India. 2015;**77**(2):170-181

[14] Griffith DA. Spatial Autocorrelation. In: International Encyclopedia of Human Geography. 2nd ed. Elsevier; 2019. pp. 308-2316. DOI: 10.1016/B978-0-08-102295-5.10596-7

[15] Roy S, Nazarkiv L, Markulchak T, Chaberek G. Identification of social economic and environmental context of

the concept of city sustainability, based on the Ternopil, Ukraine example. SSRN. 2023. Journal of Geography, Politics and Society. 2023;**13**(4):26–39. DOI: 10.26881/jpgs.2023.4.03

[16] Hristić ND, Stefanović N, Milijić S. Danube river cruises as a strategy for representing historical heritage and developing cultural tourism in Serbia. Sustainability (Switzerland). 2020;**12** (24):10297. DOI: 10.3390/su122410297

[17] Vujko A, Plavša J, Ostojić N. Impact of the "Danube cycling route" on the development of cycling tourism in Serbia. PJST. 2013;**20**(3). DOI: 10.2478/pjst-2013-0021

[18] Trišić I, Štetić S, Privitera D. The importance of nature-based tourism for sustainable development—A report from the selected biosphere reserve. Journal of the Geographical Institute Jovan Cvijic SASA. 2021;**71**(2):203-209. DOI: 10.2298/IJGI2102203T

[19] ESPON project TRACC. 2013, Applied Research 2013/1/10, Draft Final Report | Version 06/02/2015, Volume 1a TRACC Executive Summary

[20] Jovanović R, Sánchez DC, Pavlović S, Devedžić M. Principios de sostenibilidaden clusters de turismo rural: Los casos del Alto y el Bajo Danubio en Serbia. Revista de Geografia Norte Grande. 2018;**2018**(70). DOI: 10.4067/S0718-34022018 000200211

[21] Pejanović R, Demirović D, Glavaš-Trbić D, Maksimović G, Tomaš-Simin M. Clusters as a factor of competitiveness of rural tourism destinations in the Danube region of the Republic of Serbia. Tourism Economics. 2017;**23**(2). DOI: 10.5367/te.2015.0509

[22] Pavluković V, Kovačić S, Stankov U. Cycling tourism on the Danube cycle route in Serbia: Residents' perspective. Eastern European Countryside. 2020;**26** (1):259-285. DOI: 10.12775/eec.2020.010

[23] Vulevic A, Castanho RA, Gómez JM, Quinta Nova L. Tendencies in land use and land cover in Serbia towards sustainable development in 1990–2018. Acadlore Transactions on Geosciences. 2022;**1**(1):43-52. DOI: 10.56578/atg010106

[24] Vulevic A. Identification of areas with significant flood risks in counties along the Danube River in Serbia and their risk assessment. Acadlore Transactions on Geosciences. 2023;**2**(1): 1-13. DOI: 10.56578/atg020101

[25] Tomić N, Marković SB, Antić A, Tešić D. Exploring the potential for geotourism development in the Danube region of Serbia. International Journal of Geoheritage and Parks. 2020;**8**(2): 123-139. DOI: 10.1016/j.ijgeop.2020.05.001

[26] UNESCO. Djerdap UNESCO Global Geopark. 2024. Available from: https://en.unesco.org/global-geoparks/djerdap